NEW GIRL ON LOUISIANA STREET

A Novel

Doug McCall

ISBN: 978-0-9885313-8-3

Published by Desk Lamp Publishing

1-615-546-8686

PROLOGUE

February, 1992

The 737 banked gradually to the north as it adjusted its path for arrival from the east into Phoenix's Sky Harbor International Airport. The plane was circling the recently expanded American Airlines hub to line up with the runway air traffic controllers had assigned it. Late afternoon, and the setting sun made viewing the sparkling desert city below through oddly shaped oval windows from the port side of the cabin somewhat uncomfortable. The passenger in the window seat squinted into the glare, then remembering the sun shades in his shirt pocket, he reached for them. He managed only a glimpse of the Arizona capital city's downtown area to the west before the Boeing jet turned once more, straightening out for what would prove to be a relatively smooth three point landing.

Then began the part of flying he hated, and one of the reasons he had seriously considered driving the five-and-a-half hours from L. A. He sat patiently waiting, since he of course had no choice, as the other passengers clogged up the aisle, jostling each other while pulling down carry-ons from overhead bins, trying hard to be polite, producing several near-misses of brain damage. The flight from LAX was not a long one, lasting barely over an hour. But the time spent circling, then taxiing, followed by the rush to be first off the plane by his fellow travelers was almost as long as the drive would've lasted; a slight exaggeration.

He actually enjoyed getting his two-year old Chrysler LeBaron convertible on the road, putting the phony glitz, glamour and all-to-real smog of Los Angeles behind him. Knowing he would be doing plenty of driving over the next two days, he had accepted his boss' offer to spring for the ticket, plus a rental at the Hertz booth inside the terminal. This was, after all, a business trip—at least as far as his boss knew.

"Make sure you pick up one of those bag phones at Hertz. Keep me in the loop about what you find," Fritz had said when making the travel arrangements. "And pick us out a good site. Somewhere in the desert. Not too close to town, but close enough if we need to make a run in."

He had answered, "Sure," to all instructions. He knew from experience that giving long replies would only inspire the Austrian transplant to add more.

"And watch your expenses. We have a tight budget to keep in mind."

"Sure."

And there definitely *would* be business conducted over the next couple of days. But more important to the solo traveler was something of a more personal nature, something much more significant in the grand scheme of his personal life, as well as the life of the person who would be joining him—the only other person in the world . . . no exaggeration there . . . who would realize the meaning of the unofficial assignment.

In fact, if the person he would hook up with in the airport terminal hadn't agreed to meet him in Phoenix, he wouldn't have had the courage to go it alone. But she had agreed, and since she had gone to the extent of rearranging her schedule, and covered the expense of traveling here, well, there was no backing off now.

Finally disengaging himself from the now nearly empty cabin, leaving through the front door of the plane, promising the friendly but weary-looking flight attendant he would "have a nice day", he edged his way through the crowd gathered in the gate area; most were de-planing, but a few were waiting for the next flight out of the gate. He carried only a thin briefcase, containing a list of places he planned to visit while performing his "official" duties. It also contained one sheet of note pad paper, with only a single address written on it. It was by far, the most crucial item in the brown valise.

Finding his way to the Delta gate where the other party would be arriving, he checked the ETA on a board in the area. A delay of some fifteen minutes gave him a chance to settle into a hard, molded plastic chair, and think through the possible outcomes of the real reason he had lobbied his boss so hard to allow him to come to Arizona. The task he had been sent to accomplish was actually someone else's job, but he saw it as a chance to satisfy his own personal agenda. It could prove to be disappointing—even a monumental disaster.

He knew, however, there would be a void in his life if he at least didn't try; a void that would last possibly forever. He had been through such a hollow period before, but if things went well, hopefully it wouldn't happen again.

So this journey was important to him, and to his soon-to-arrive companion. There was history involved, history that involved them both. It had all begun with that awkward, strange new girl on Louisiana Street, and her equally odd family. How much they had changed his life—his outlook on life. So the reason for coming here involved something more than simply being important; It was amazing. In fact, it was more than amazing; it was unbelievable.

CHAPTER ONE

The welcome surprise of an early October chill had settled in over his part of Texas, one which the citizens of the area knew wouldn't last. It was only a tease for the natives of the northeast quadrant of the Lone Star State, lingering for only a few days before a few parting blasts of super heated conditions returned; after all, this was Texas, not known as a temperate state. But it would eventually, mercifully, give way to the usual tolerable fall weather that would arrive later to stay.

The town of Cypress Springs, population just north of 8,000, was a nondescript community with no particular characteristics that made it stand out, even among the similar communities of like size in the region. Nestled within that corner of Texas which served as a hinge of sorts, both connecting and separating the southeastern portion of the U. S., from the area referred to as the southwest, although it wasn't really "The Southwest" people normally pictured in their minds.

Cypress Springs had been given its name due to nearby Cypress Lake, which was supposedly fed by a spring, even though no one had ever located the source that lead to the meandering creek, and finally to the bucolic 500 acre lake. It was a town you were either born in, or transferred to, usually against your wishes. It would never be seen as a destination spot.

Positioned within shouting distance of the so-called Ark-La-Tex region, it possessed features of all three states: Cajun dialects and spicy recipes brought from Louisiana, southern drawls and barbeque from Arkansas, and the natural boastful pride of actually residing within the boundaries of Texas. Chicken fried steak and Tex-Mex ruled here, along with any other dishes that could be classified as so-called "comfort foods". The pines of Louisiana and Deep East Texas began to thin out here, giving way to elm, oaks, cypress and a few ash tree varieties for the next few hundred miles west, until the "real Texas" appeared, with mesquite brush, oil wells, longhorns, and wide-open spaces.

The surrounding countryside, surprisingly hilly in some portions, leftovers from the Ouachita Mountain Range that had receded eons ago, leaving behind mostly gently rolling terrain representing the northern slice of East Texas.

Aside from those factors, the region was virtually unknown to the rest of the country. The majority of the local citizens likely preferred to keep any outside knowledge at bay; not because there were secrets to reveal – well, maybe there were a few, but even some of those were sources of pride. It was more a case of the belief they didn't require anyone's approval of their way of life.

Mickey Foreman possessed no personal attitudes toward any of these demographics or social mores of his surroundings. At age eleven, he saw it as a challenge just to live life one day at a time, with the possible hope he clung to that at some point in the future, he would leave Cypress Springs far behind.

The year was 1978, and things were changing rapidly in the world outside northeast Texas; things he wanted to be a part of. It was now possible to watch movies any time you desired, with an amazing device called a VCR, short for video cassette recorder, something so incredible to Mickey, it seemed magical. In fact, since his small family couldn't afford one, it *was* still unattainable magic to him. If he'd known his dreams were the same as a million other boys his age, it wouldn't have mattered to him; he would try his best to follow his.

Fortunately for Mickey, the family of his best friend, Benny Payne, actually possessed a VCR, which allowed him occasions to view the types of movies that were important to him. Films, such as *Logan's Run*, *Close Encounters of the Third Kind*, and especially *Star Wars*, thrilled him, leaving him restless for his future. All too infrequent trips into Texarkana and nearby Shreveport, Louisiana, with his beleaguered mom and sister to view movies in newly built theaters, were special treats.

On this otherwise agreeable afternoon, there was enough of a chill in the faint breeze to cause him to zip up his windbreaker and slide his hands deep into his jacket pockets. The shiver working through his body had little to do with the elements however. He had just seen something, only minutes ago, that had his nerves on edge. Confusion, laced with an undeniable touch of fear, had overtaken him. He had never seen anything like it—at least, not in real life—maybe in a movie scene that he probably shouldn't have been allowed to see. And he was sure he wasn't meant to witness what he had; it was only by accident that he had—a sudden detour in his path had made it possible—a detour he now wished he hadn't taken. But he had, and now he had to somehow find a way to erase what he had seen from his thoughts.

He had been navigating his way home from the Dairy Queen on the edge of downtown, taking a familiar route. The DQ was a favorite place to hang out after school, giving Mickey time to spend with classmate, Kristi Shields. He had never even touched the blue-eyed girl, who always smiled when she saw him. Her straw colored hair hung almost to her shoulders, swaying when she laughed or shook her head. He wasn't brave or foolish enough to label her as his girl friend, at least not out loud, feeling they were both too young for that classification. But she was a lone beacon of happiness for him, someone who made the entire town light up in his eyes.

They would simply sit across from each other in a booth, drinking whatever Kristi's older sister, Karla, a part time employee, could comp them when the manager wasn't looking. If he *was* watching, they would buy the cheapest soft drink on the menu board. Karla was sixteen, five years older than her sister, not bad looking herself. But Mickey, more than anyone, knew his limitations. Plus, she had a boy friend who played on the high school football team, not to be messed with.

Between long periods of saying nothing, the two fifth-graders would gossip about fellow students, or discuss tv shows seen the night before, until either of the two felt it was time to head home for supper. This was the full extent of their blazing romance.

To save time on this brisk afternoon, Mickey had decided to take a short cut, leaving the sidewalk to walk across backyards of a couple of familiar homes in his neighborhood, as he had done many times. Stepping over a low hedge bordering two houses, he glanced into the kitchen of one of the homes, as he had also done many times. In fact, he had been in that house several times over the past few years, even in that very room. Peering through the large window of the kitchen, he saw movement.

His first thought was that someone, probably the woman of the house, was preparing something on the circular table he had seen in the room before. His second thought, after looking closer, was that the woman was naked. Whoa! He could see a bare bottom leaning over the table, a sight that immediately grossed him out. Maybe it was his sometimes friend who lived there, tussling naked with someone, a thought even harder to stomach.

Like a car wreck spotted on a highway, he wanted to look away, but couldn't make himself stop staring. His next recognition made it even tougher to avert his eyes; the bare bottom belonged to a man—an adult! And he was leaning over a figure on the table, a figure that he could now see was a female, whose pale white dimply rear end was also visible. Knowing the female had to be the woman he had seen in the house before had him transfixed, unable to move his feet, even with the knowledge he may go blind by continuing to stare. He knew the man was not her husband, because *he* was definitely *not bald*, as this man was. And he knew for a fact her husband was not in town. He was on the road down south.

As the man began to straighten up, he was tugging on his pants. He then walked out of view. His identity, as he stood up. had suddenly became all-too clear to Mickey. Managing to put his legs in motion, the boy stumbled, fell face-first, gathered himself, and ran.

CHAPTER TWO

Wanting to separate himself from the troubling scene he had just observed, Mickey wasted no time circling the next house in line, and seeking the relative safety of the sidewalk on Louisiana Street once again. Feeling the uneven, cracked concrete under his shaky feet, he looked around for witnesses. He wished it had been darker, then he would feel better about his chances of not being spotted by anyone; especially by the man in that kitchen, doing whatever it was he was doing to the woman. As it was, he wasn't completely certain he hadn't been.

His adrenalin flowing, the chilly conditions ceased being a major factor any longer. His mind was jumbled, trying to decide what he should do, if anything. What could he do? Who could he tell? And what would he tell them? And if what he saw was what his imagination told him it was, his know-it-all friends had some explaining to do.

He knew without being told that boys his age were always prone to exaggeration, giving the impression they had knowledge about things of a sexual nature, while having no experience to back up their claims. It was their nature to seem worldly beyond their years. According to his sister, Janice, two years his senior, girls did the same. But, as she explained, there were different motives for each gender. Boys imparted their supposed wisdom in order to create an advantage for themselves when their ages grew to the stage when actual opportunities matched up with increasing sex drives. Girls, on the other hand, used knowledge shared by their peers to *prevent* boys from using this advantage until *they* were good and ready for them to do so. Mickey couldn't wait to let his friends know they were full of it. He knew more than they did now.

One thing he knew for sure; he had to keep the effect of what he had just seen from his face when facing his mother. Besides, it would not be an easy thing to discuss with a parent. This was not the usual topic of conversation while sitting down to Sloppy Joes and tater tots, or whatever she was going to put on the table.

His thoughts were interrupted by a cheerful voice, a voice seeming to float through the air, hailing him from the front porch of a house on the opposite side of Louisiana Street as he continued down the sidewalk. It was a voice he was familiar with; a pleasant one

"Hi, Mickey," she said.

"Oh, 'lo, Heidi," he answered, surprised to see her outside. Her tone was soft, even while projected across the street. It seemed to have strength behind it though, carrying her voice through the wind, as it always did in the classes they shared.

"This is pretty cold for Texas, isn't it?" she said.

"Well, it gets this way sometimes. It'll get hot again soon enough."

"I'm sure you're right, but it feels pretty cold now."

Heidi Jones was new to Cypress Springs, having arrived with her family late the previous spring, moving in from places unknown. Tall and thin, with hair so blond it was practically white, seemingly transparent.

She had a calm nature that made it easy to like her, but Heidi remained somewhat of a mystery, not only to him, but his other classmates as well.

"Been to see Kristi?" she said with a visible grin, hands clasped in front of her trim body.

Wondering how the heck she knew that, not that it was the subject of his confusion at the moment, he answered, "Yeah." He started to continue on, then not wanting to be rude, added, "How're you doing?"

"Okay." She then wrapped her slender arms around her shoulders.

Glancing at the window to the right of the girl, he saw someone peek out, before quickly disappearing. He knew nothing of her family, only that she lived with them in the house that had belonged to old Mrs. Frederick until she had died a couple of years earlier. As far as he knew, it had remained vacant until Heidi's family took possession in either April, or maybe May; he wasn't sure which. They were just suddenly there.

"Guess I'd better get home," Mickey said, stepping sideways. "See you tomorrow, Heidi."

"Okay, Mickey. Good night." Unwrapping her arms, she turned toward the door.

Two more houses to the next corner, then a right turn to the next corner, and he was home. He paused on the second step of the three bricked steps leading up to the small concrete landing in front of the door. It was important to wipe all tell-tale expressions from his face, or else face the third degree from his inquisitive mom. She had enough to deal with without uncovering her son's accidental glimpse into what he was sure was some form of adult depravity.

Jessica Foreman was the unfortunate widow of Hampton Foreman, better known as Hamp. A Marine First Lieutenant, he had been one of the latter casualties while serving in Vietnam. The unpopular war was supposedly in mop-up mode in early 1971, during his second tour. However, it had been impossible to account for all of the land mines buried over the years of the encounter, either by the North Vietnamese, or their benefactors, Communist China. After Lieutenant Foreman and two of his men happened onto one of the devices, killing him instantly, and horrifically injuring the other two, there was one less to worry about locating. There had been some life insurance payable, but not nearly enough for the young widow, now working as the secretary to the owner of a small construction company in neighboring Mt. Pleasant, requiring a daily commute of twenty miles, each way. The job was stressful, primarily due to the fact commercial construction was by its nature stressful. Her boss did his best to be tolerant of the domestic duties of a single mother, but he had his difficult moments. The pay was decent, and while she grew to dread the morning drive to work, the trips home through the wooded countryside allowed time to decompress, so Jessie, the shorter name she preferred, was resigned to her fate.

Since it was normally closer to six than five, and often later, when Jessie arrived home, it gave Mickey free time to wander the nearby neighborhoods, flirt with Kristi, or hang with his friends. His older sister, Janice, mature for a girl of thirteen, had taken on the role of managing the household between the time school let out and their mother arrived.

"Hi, honey," Jessie Foreman said as Mickey slammed the front door behind him a little more forcefully than intended. He was still on edge, in spite of his effort to steel himself.

"Hi, Mom. Sorry I'm a little late." He didn't make eye contact as he walked past her.

She was standing in thc short hallway, just having placed her purse on the marble-covered hall table while picking up the mail Janice had collected. She kicked her shoes off. "I just got home myself, so no biggie. Got a new bid accepted today. Mr. Montgomery wanted to go over it with the staff."

"Cool." Stepping down the hall, he took a quick glance back at his mother, now poring over the mail, mostly of the junk variety. Only thirty-five, she looked closer to forty. Grieving, followed by a life she hadn't planned for herself, had taken its toll. Worry lines were starting to form around her eyes; her brownish hair seemed permanently disheveled, with a few specks of beginning gray. But her somewhat scruffy appearance gave her a carefree look that only added to her appeal. Her eyes, like her daughter's, were a light shade of gray. Mickey's colorings, including brown eyes, came from his dad. Regardless of the stress she faced daily, he thought his mother was one of the prettiest women he had ever seen. With her looks, he often wondered if she would remarry, bringing a new parent into their lives. He had several classmates who had step-parents, which more likely than not proved to be disastrous, at least from their perspective. There had been callers the last couple of years, including her boss, who was divorced. He seemed okay, but with no kids of his own, he usually appeared to be on edge around his secretary's two offspring. Anyway, that situation seemed to be dying out, thankfully.

"Where's Jan?" he asked, hoping to do his best not to betray signs of what he'd seen only a few minutes earlier.

"In her room. Think she's studying."

Mickey walked nonchalantly past the modest den/living room, taking a right at the end of the hall, down a shorter hallway. Her door was cracked open, so he knew she wasn't doing any heavy studying. He could hear the moody sounds of Janis Ian emanating from the tiny bedroom. Similar to his room across from hers, it included a twin bed, a small wooden desk and matching chair, a small chest of drawers, plus a cramped closet in a corner. She was sprawled on the bed, with a book actually open, listening to the Panasonic cassette player sitting on the chair. The song, *At Seventeen*, the singer's one-and-only popular hit, was playing. It was his sister's anthem, and her almost namesake was her favorite singer in spite of her introverted style and lack of star power.

"Okay to come in?" Mickey asked through the opening.

She answered without looking up. "Of course, my good man. Just try not to make an utterance until this song's over."

He noticed the book had something do to with English literature, so that explained the haughty attitude and language. Since the only chair was occupied by the cassette player, he inched his way onto a side of the bed. She slid over to give him room.

He waited as directed. It was in fact a good song, though a little depressing if you actually listened to the words. Ian just couldn't seem to break through to the mainstream. Even a couple of shots on early *Saturday Night Live* broadcasts hadn't done it for her. Just too sad looking in Mickey's opinion, but he would never voice that to his sister. He didn't believe his sister saw herself as an outcast, ugly duckling teen, as in the lyrics of the song. In fact, her looks were better than average, but chose not to use what beauty she had. Slightly freckled, with similar hair to Jessica's, cut short, she was tall for her age, which her slender build emphasized.

The song over, Janice reached to turn the volume down on the player, closed the book and sat up to prop herself against the headboard.

“What’s up, Mick?”

Now that he had entered her domain, he wasn’t sure why he had; he wasn’t sure what to say. He settled on, “Mom was busy, so thought I’d see how things are with you.”

She stared with a frown. “All of a sudden, you wanna know how I’m doing?” Always blunt, she never hesitated to get right to the point.

“No. Not all of a sudden. I always like to know what you’re up to.” Even to his ears, it was lame. “Just don’t take the time to ask.”

Obviously not believing him, she evidently decided not to make an issue of it. She stretched and sighed. “Pretty good I’d say. Liking my classes so far. Good to get back in the swing of school, see my friends, keeping my grades up.”

He pursed his lips and nodded. “That’s good. Any new boy friends chasin’ you?” He laughed nervously as he asked, aware his obvious poker face was evaporating.

She stared. “Boy friends? All right, little brother, what’s going on? Puppy love with Kristi not working out? Either spill it or let me get back to my reading and my Janis.”

He knew then he wanted to tell her; in fact, needed to tell her. Get it off his chest. She wouldn’t freak out like his mom would. Without speaking, he went to the door, closing it as quietly as he could, then turned to stare at his sister for a second or two.

CHAPTER THREE

Mrs. Robinson! Are you sure?" Thoroughly shocked by her brother's story, Janice couldn't keep a grin from bubbling up.

"Course I'm sure. What other woman would be in the kitchen of *her own* house?"

The siblings fell silent for a few seconds; Mickey looking at Janice; Janice staring through the horizontal window high on the wall. Her gaze was in the general direction of the Robinson house, as if she could see the carnal activity her brother had seen from her own bed.

"And you're certain the man wasn't her hubby?"

He shook his head rapidly. "Yeah, I'm positive. Last time I saw him, Mr. Robinson had hair. Plus, Butch told me earlier this week his dad was working a job in Houston. Be gone a couple of weeks."

"Maybe he came home early."

Another shake of sandy hair, followed by a heavy sigh. "Jan, I know who he was. I saw him as he—you know—finished, then stood up."

She leaned in. "Who?"

Knowing the man's identity himself was bad enough, but telling Janice might get her in trouble too. He knew if he refused to tell her she would get mad, but if he told her the truth, that the man was known to him as

Chief Winston Dunaway, Chief of Police of Cypress Springs, Texas, it would put her in the same bind he was in. "You gotta trust me, Jan. I can't tell you. We'd *both* be in a heap of trouble if you knew."

Other than squeezing her eyebrows together, she didn't react right away. He took advantage and added, "Besides, does it really matter who it was? Crap I just wish I hadn't cut through the Robinson's back yard, hadn't seen it at all."

Janice seemed to accept her brother's anguish, choosing not to push him. She fell back again on the headboard. "You sure nobody saw you?"

"Don't think so. I didn't see anybody, and I'm pretty sure *they* didn't see me." He thought for a second, then added, "Only person I saw was when I got back on the sidewalk. Heidi Jones was on her porch. She yelled over to me."

"You mean that freaky new girl? She's weird."

"She's not that bad. I kinda like her. Quiet, smart. Besides, you've never even met her. You're in middle school now."

"I know, smart aleck. But my friend Barbara Sue Wilson's little brother Jake's in some of her classes. He says she's a flake."

Mickey snorted. "Well, Jake oughta know about flakes. He's the biggest I know."

They grew quiet again. Realizing he still had his jacket on, Mickey tugged it off, tossing it on the floor. He thought once more about what his friends had said.

"Do you think what I saw is normal? I mean, is that what adults do when they—you know—do it?"

Janice considered the question for a moment. "I guess, sometimes.

Amber Dunn told Barbara Sue and me she snuck a look at a movie on HBO when her parents were out. She saw a couple in the film screwin' in a way that sounds like what you saw."

Mickey couldn't disguise the shock at the descriptive word his sister used. She was growing up a lot faster than he was. She noticed it. "Oh, come on. That's not such a bad word. You'll hear much worse when you get to middle."

"Maybe so." That warning both thrilled and scared him at the same time. "But I can't wait to see Benny and Trey tomorrow and tell them they don't know everything they thought they did."

"What about Butch? You going to tell him you saw his mom doing that?"

"Are you crazy? He'd call me a liar, tell me to shut up. Then he'd probably beat the crap outta me."

Butch Robinson wasn't an official member of his small group of friends, which included Benny Payne, Trey Fulghum, and Roby Contreras, a relative newcomer they had let in primarily due to the fact his family owned the best Tex-Mex restaurant in town. He was a good guy, and fun to hang with. Butch, on the other hand, was a bully, who no one liked. His tough guy attitude alienated most everybody, but on occasion he would weasel his way in to Mickey's group.

A knock on the door actually made Mickey jump. Jessica Foreman opened the door, and announced, "I'm too tired to cook tonight. About to order pizza delivery. Any requests?"

"Anything but anchovies and mushrooms," Janice said.

"Pepperoni and whatever," Mickey added, somewhat subdued.

Jessica looked from son to daughter. “Everything okay?” she asked, her worry antennae activated. Sequestering themselves in Jan’s room wasn’t a normal occurrence.

“Yeah,” Janice answered. “Just talk. School stuff.”

Doubt showing, the tired woman nodded and closed the door. They waited until they could hear her making the call, then continued.

“So what’re you gonna do?” Janice asked. “You know, I’m not completely sure about this, but I don’t think there’s any crime involved in what you saw, except maybe crimes against nature. Oh, and for you being a peeping Tom.” She teased with a grin as she added that.

“Funny girl. I just did what I usually do, took a look to see what was going on.” Saying that, he knew it did sound bad. “There’s nothing *to* do. I’m just gonna try to forget it.”

“Good luck with all that. You sure you don’t want to tell me who the man was?”

“Not now. Maybe later.” *But probably never, unless he had to.*

CHAPTER FOUR

Over the course of the last six or seven years, the time during which Mickey Foreman's brain began to form an understanding of things occurring in his surroundings, and what they meant, there had been only a few events that stuck in his mind as being worthy of remembering with any clarity.

The first was the fire, occurring five years earlier, torching three buildings on the town square: a hardware store, insurance agency, and a decades-old diner, all in a straight line on one side of the square. Most folks were quick to blame the inferno on a grease fire in the tiny restaurant, a claim that was never proven, and that the incensed owner denied. But it made more sense than anything else. Regardless, the stench of soot, along with incinerated lawn and garden chemicals from the hardware store, wafted through the entire town for days. Gas masks were needed in the downtown area.

Next came the tornado two years later, skimming along the northern edge of town. Weather officials that ranked these terrors wrought by Mother Nature gave the twister a classification of F-2, which was a huge disappointment to the citizens of Cypress Springs. They were hoping for a F-4, or at the very least a F-3. It had something to do with Texas pride, wanting nothing but the biggest, but mainly it would've made a difference on insurance claims and disaster relief. The swath left across a few low-end neighborhoods, businesses, and trailer parks produced three deaths and clean up that lasted for months.

These two catastrophes stayed with the locals for an extended period, longer than the time needed to put things right. The third event earlier in the current year, while seeming spectacular to Mickey and his friends, was more-or-less forgotten within a few days of its occurrence, and not even noticed by most citizens. It had lasted over two nights in mid-March, as far as anyone could recall. Bright lights streaking across the Lone Star sky at night, visible from near the Oklahoma border to the north, to within ten miles of I-20 to the south, a distance of some sixty miles. There were even sightings reported from across the state lines of Arkansas and Louisiana to the east. Every observer agreed the ill-defined trajectory of the trail was generally from north to south. Meteor showers were the most popular answers, even though there were no craggy rocks or craters found. If they were indeed meteors, they had either burned-out before reaching Earth's surface, or likely been deposited in one of the many area lakes or ponds, often referred to as "tanks" in rural Texas.

Mickey and his gang of four had been awe-struck; what could be cooler than objects from space flaming through the star-lit skies above, or better yet, actually reaching ground. They had begged their respective parents to help launch a search, an idea that was quickly nixed, since no one had any idea where to start. "*If the experts can't find them, what makes you think you can?*" was the crushing comment from Benny Payne's dad. In any case, those few days and nights of wonder had folded well into the backs of their memory banks by mid-summer.

The bombshell Mickey laid out for Benny and Trey Friday during lunch period was now added to the list; a more personal event in their young lives, and something entirely different. Roby Contreras was absent, having taken advantage of a rolling excused absence he used on occasion when his help was needed at his family's restaurant to prepare for the Friday night rush. Following the last morning class, which was civics, he was allowed to depart. He was always glad to skip out early, but the work waiting for him could be grueling, with any potential pay depending on the night's receipts, the pay being treated as an allowance to avoid child labor laws.

It was just as well, as far as Mickey was concerned; the fewer ears the better. In fact, he would've been happier with only Benny, the friend he knew he could trust the most. To play it safe, he decided on his way to school that morning to omit names from his expose, which he knew wouldn't go over well. The three remaining friends sat at a metal picnic-style table on the concrete patio-like surface immediately outside the lunchroom, trying to digest the mediocre tacos and beans they'd just scoffed down, while Mickey told his shocking story. The rare morning chill had given way to mid-eighty temperatures, but the air outside felt good.

"Come on, Mick. Tell us who they were," Trey Fulghum pleaded. "You can't tell us all this and not give us names."

Mickey knew at once he hadn't thought things all the way through; Trey could be persistent, like most red heads. His freckled face was now flushed with curiosity, making his whole appearance look even more crimson. Both Benny and Mickey had been full-on blond when they'd first met in first grade. Over the years, while his had turned sandy, with ever-present bangs, Payne's had darkened to a nearly brown shoe polish shade. Contreras, when he was with the group, added a touch of brown skin to the group, which offset the other three.

"Trey, what difference does it make? I don't want to get things stirred-up in school. Only wanted you idiots to know what y'all told me was wrong."

"What was wrong? What're you talking about?" Benny asked.

Mickey gave his best friend a sarcastic grin. "What you and Fulghum told me about sex, and how it's done. You never said anything about doing it that way."

"Christ, Mick! We didn't mean the other way was the *only* way to do it. Just the normal way."

"Yeah, that's right," Trey added, shaking his head. He squinted and said in a voice that indicated he was sharing the deepest of deep secrets. "Heck, I'll bet there are a few other ways we haven't even talked about yet."

Mickey thought about their answers for a few seconds. Were they still trying to come off as smarter than him in this particular area, or just blowing more smoke? Either way, the conversation was quickly drifting off the rails, in a direction he didn't want to go. But, too late now.

"Hey," Trey said suddenly, looking at Benny. "Bet we can probably figure out who it was—the woman anyway."

"How?" Payne asked, as Mickey's stomach began to knot up.

"Well, Mick said he was walking home from the D Q, cuttin' through back yards. We can figure out which houses he would've gone behind between those two places."

"You mean, sorta like detectives do? like Columbo?"

"Yep. It'll be easy." Fulghum gave Mickey an evil grin.

Things were really out of hand now. Mickey knew he was showing panic on his face. Just then, the situation gave signs of developing into a perfect storm, as Butch Robinson came loping out of the lunchroom, digging into a paper cup of ice cream with a thin, flat wooden spoon. Spotting the trio, he headed straight for the conspirators huddled at the bench.

Crap! Mickey knew if the two amateur sleuths solved the puzzle before he reached them, they could very well share the knowledge with the guy who would then proceed to kick his butt into tomorrow. He had to squelch things—and quick.

Deliverance came from seemingly nowhere, in the form of a slight girl whose approach toward the bench from another exit from the lunchroom could only be described as gliding over the concrete. Her feet hardly seemed to touch ground. It was Heidi Jones, smiling broadly at the boys at the bench, but mainly in Foreman's direction.

"Hi, Mickey. Guess you made it home okay last night?" Her light features were even more pronounced up close, her skin appearing almost transparent, with green eyes under a high forehead covered by that weird white hair. The voice was the same; soft yet strong.

Benny and Trey looked at the girl, then stared at the third member seated at the table with curiosity and amusement.

"Yeah, sure, Heidi. No sweat," Mickey said, attempting a smile that came out crooked. "It's not that far from your house to mine." Her comment was an innocent one, but he was hoping to use it to shift attention off of the subject.

Butch finally lumbered up to the table, chocolate ice cream spread across his mouth. Half a head taller than the other boys, in fact taller than everyone in the fifth grade due to being held over one grade, he wore his clothes in his usual sloppy manner over his chunky body. Having no idea what the other boys had been discussing, and not having been a part of it, he still chose to throw his weight around.

Frowning at the girl, he said, "Whatta you want, you creepy scare-crow? This is our meeting. Get lost."

"Butch!" Mickey said. "Why do you have to be such a jerk?"

"Yeah," Benny threw in. "What's wrong with you, Robinson?"

"What?" Butch replied, wiping his mouth with his shirttail. "You tryin' to turn this into a tea party?" He sneered at the girl. "Get lost."

Mickey started to rise from his bench seat. "Butch . . ."

"Don't worry about it, Mickey," Heidi said, showing no fear or embarrassment in her voice. She smiled his way again, then headed for the building.

Mickey was about to say more when the bell signaling the end of lunch hour sounded. He, Benny, and Trey then hurried inside. As they did, Trey said, "We'll talk about this later."

Emitting a loud belch, followed by what had often been labeled by his classmates as bazooka farts, Butch followed. "Talk about what later?" he asked.

CHAPTER FIVE

Moving up to the fifth grade had put Mickey and his classmates in a new world, classroom-wise. Where they had spent the entire school day in one class with the same teacher before, they now had separate rooms for each subject. The experimental process had been adopted recently by some local Texas school systems to prepare students for the type of instruction they would face in middle school and beyond. Mickey enjoyed the change, feeling it broke the school day up into manageable parts. Plus, it made him feel older, which was always a good thing for an eleven year old.

The first class following lunch was biology, where he endured the yawn-fest of learning about the types and structures of leaves. Then halfway through English class, a sickening thought began to creep in, as events that could easily transpire entered his mind. Sinking low into his desk, he could picture how the future might play out. One of his friends—most likely Trey—would pass along what he had been told at lunch. Stories like this would spread like wildfire through the school, then into the community. The chief of police, when it reached his ears, wouldn't waste a second trying to protect his image. Whether his name was mentioned or not, he couldn't take the chance it hadn't been. He would hunt down whoever had seen his involvement. Mickey could already feel the cuffs on his thin wrists.

"Oh crap," he muttered, not realizing it had left his mouth loud enough to be heard. The students within a few rows all turned to look. Mrs. Crawford, the teacher, stopped in mid-sentence, turning to look in his direction.

"Mickey, do you have something to share?"

He sat up quickly. "No, ma'am. Just thought of something I forgot to do."

Mickey Foreman was not her star student, but had never caused problems either, so she let him off easy. "Well, I trust it wasn't your homework assignment," she said with a sardonic grin.

"No, ma'am," he said sheepishly to a chorus of muted laughs.

With his mind still occupied by thoughts of the police chief, he made it through English, moving on to his final class of the day, math. He had done okay during the last couple of classes with computing decimals, but the steps involved with converting fractions and reducing them to the lowest common denominator were giving him fits. Today especially, with his brain still cluttered with images of Chief Dunaway breaking his front door in to arrest him—or worse.

Glancing around the room to clear his mind, he noticed Heidi Jones directing one of her enigmatic, gentle grins in his direction. It may have been his imagination, but she seemed to be recognizing his distress. He quickly looked back at his desktop. Someone else saw the girl's attention levied toward Mickey. Kristi Shields, sitting three rows behind, centered between Mickey and Heidi, frowned as she took note, before focusing again on her work.

Later, as the bell sounded, ending one of the longest school days Mickey could recall, he turned down the hall to begin his long trudge to his locker to collect his book bag before catching his bus ride home, when he was stopped by Heidi's unique voice. She approached.

"Mickey, thanks again for speaking up for me at lunch, although it wasn't necessary."

"No big deal. Butch can be a real pain sometimes." He was turned sideways to the girl, inching away, anxious to catch the bus before being left behind.

"Yes, I've heard that." She paused for a couple of beats. "Listen, why don't you let me help you with those fractions. Looks like they're causing you a little trouble."

More than a little. "That's okay, Heidi, I'll get it eventually."

"I'm sure you will. But if you can come by my house, maybe this afternoon, I can have you mastering them within fifteen minutes. Then you can impress everybody with your skills Monday."

When he paused for a second, she continued. "Please, Mickey, let me do something for you. You can come by anytime. My family is very informal. Just knock on the door."

He was about to refuse again, but from her almost pleading tone, he could see she probably needed a friend. No one had warmed up to her at school, at least that he knew of.

"Sure. What time?"

"Anytime is fine. But we can say, maybe five thirty."

"Okay. Sounds good."

"That will be fine. See you then." Backing away, she added, "Thanks, Mickey."

As he walked rapidly down the hall toward his locker, he didn't look back. If he had, he would've seen Kristi, watching from the far exit door, with a questioning look on her face.

CHAPTER SIX

Arriving home thirty minutes later, Mickey stopped off long enough to deposit his bag on his bed, then prepared to head to the D Q. Janice stopped him as she entered, meeting him at the front door.

"Hi, shrimp," she greeted, using the nick name she knew her kid brother hated. "Any news on you-know-what?"

She could only mean what he had seen the day before. "Nope. Nothing yet." Even though he had told her he was going to confront his friends about their misinformation, he decided to keep the lunchtime meeting to himself. Plus, he didn't want to get into a long thing with her, anxious to meet up with Kristi.

"Gotta go. See you later," he said as he opened the door.

"Say hi to Kristi," she said, grinning as he went out. "And don't forget Mom might want to go out tonight, so don't be late." Her last words faded away as he closed the door behind him without a reply.

The chill of the previous day now completely evaporated, Mickey was glad to enter the familiar restaurant, leaving the relative furnace outside. He was quick to learn, however, that he was walking into an icy frost zone. Kristi sat in their usual booth, but her body language was anything but normal. Not looking up at all as he slid onto his seat across, she barely mumbled "hi". There was no drink in front of her on the table; not on her side or his. She usually had them ready if Karla had been able to sneak them to the young lovebirds. Starting to feel confused, and a little uneasy, Mickey looked at the girl, who was still staring at the table top. Glancing at the counter, he saw the big sister handing over soft-serve ice cream cones to a couple of customers. Taking payment, she then peered over at the young pair, with a furrowed brow showing a hint of concern brewing. He made several attempts to pull something out of the same girl who had laughed and teased him only the day before; the same girl he had known since first grade, performed in school plays with, and seen around town for years before finally striking up whatever it was they now had. Only on this particular afternoon, she was reacting as if they were complete strangers, perhaps even enemies.

This was something new for the suddenly lost fifth grader. He had no reference points, no history to call on to explain her behavior. It must've been something he'd done—but what? Thinking back over the past twenty-four hours, since they had parted on what he was sure were good terms the day before, nothing came to mind. He couldn't see any way she could be aware of what he'd witnessed the day before behind Butch Robinson's house. If she somehow knew about it, she probably wouldn't be here at all, feeling it was too creepy.

After a few more attempts, receiving only muted, terse replies, he gave up. Saying "I'll see you tomorrow" in a tone he was hoping showed his confusion, he slid from the booth seat.

Noticing him leaving, Karla yelled over from the counter, "Take it easy, Mick."

He nodded without saying anything, glad that at least one of the Shields sisters was speaking to him.

Trudging slowly across the parking lot, he didn't look back. Again, if he had, he might've seen Kristi's eyes following him until he vanished from sight. Karla stepped out from behind the counter, going to check on her younger sister, who now had her head in her hands, her eyes beginning to well-up.

Feeling bummed out over the way things had gone, or actually not gone, with Kristi, Mickey was relieved to find a note left by Jan when he returned to his house. She had gone to a friend's house, and would return by five o'clock. Which meant he wouldn't have to try and explain why his trip to the D Q had been so short. Which also meant he had time on his hands before heading to Heidi's house, a promise he had almost forgotten. Still not quite four-thirty, he grabbed a soft drink from the fridge, along with a half-empty bag of chips, and settled in the den to watch a re-run of a *Leave it to Beaver* show. He kept running the experience with Kristi through his mind, trying to figure out its meaning. Maybe this was how these things went; lasting a few months before one of the participants moved on to something, or somebody else. It was all just too new to him. Maybe he would ask Jan about it later. But then again, maybe not.

Heidi had suggested five-thirty as a time to get together, but she had also said any time would be fine. So, when the show ended at five, he tossed the drink can in the garbage can, put the chips bag back on the kitchen counter, grabbed his math text book, and began the three minute walk to the Jones' house.

Crossing the street at the corner of Sabine and Louisiana Streets, staying on the side opposite the Robinson's house, he quickly reached Heidi's house. He looked the structure over as he hadn't done before, never having reason to call on Mrs. Frederick. It was on the plain side, not that that meant anything to him, with aged red brick across the front. There was a landing similar to his house, but without an overhang. The two large front windows on each side of the door had curtains pulled shut within. He approached the front door, having no idea it would be the first of many times.

Unable to locate a doorbell, he was preparing to knock when the door suddenly opened, and the girl stood there, evidently aware of his arrival.

"Hi, Mickey," she greeted. "Glad you could come by."

"Sure. Hope it's okay I came early."

"Of course. Please come in."

He went in cautiously, entering the modest home for the first time. If he'd had an inkling of what lay ahead, of the many times over the coming days and weeks he would walk through that door, and interact with the unique characters residing here, he would've had an entirely different feeling. But there was no hint of what was ahead; no way to predict how his life would be changed in ways he never thought possible.

CHAPTER SEVEN

There was no entry foyer or welcoming hallway, just a large rectangular room serving as the living room/dining room combination. For a room of its size, there was a minimal amount of furniture: a long sofa of some golden fabric backed up against one of the windows; two cloth wingback chairs on the opposite wall, separated by a lone night stand, on top of which sat a lamp with crystal-looking teardrop beads hanging from the shade. An oval dining table sat at the far end, with four straight back chairs encircling it.

An aged green shag carpet ran almost the length of the room, but ended just short of the dining area, where yellowed linoleum could be seen. Two doors exited the room, one on each end, evidently emptying into the kitchen, plus any other rooms that might be hidden from view back there.

"We can use the table to work on," Heidi said, turning in that direction.

He followed without comment. The girl was all business, not wanting to waste any time. It fit with the efficiency he had observed in the classes he shared with her. As always, she seemed to glide across the floor, her feet barely grazing the tufts of the carpet.

As they took their seats at the wooden table, a woman appeared through the near door, coming to greet him. Assuming this was the mother, Mickey saw her as a taller version of Heidi, also on the thin side. The primary difference was instead of the girl's nearly white, fine straight hair, the woman had what was apparently a brownish hair piece perched awkwardly askew on her head.

"You must be Mickey," she said, as she stopped behind Heidi's chair.

"Yes ma'am. Mickey Foreman." He didn't stand, but rose slightly in his seat.

"Please, Mickey," the woman smiled and replied. "It's Blanche. Heidi may have told you, we're not very formal here."

"Oh, yes ma'am . . . I mean, Blanche."

She smiled again. "Is Mickey Foreman your full name, or is there a middle one?"

"Yeah. It's actually Mitchell Knox Foreman. Knox was my mom's maiden name. I've been called Mickey all my life."

"Well, Mickey, we're so glad you came by. Heidi's told us so much about you. Please feel free to visit as often as you like."

Somewhat curious how the girl knew "so much" about him, he simply nodded, and said, "Thanks".

"Okay then, I'll leave you two to your work. If I can get you anything, let me know."

"Thanks, Blanche," Heidi said, as the woman went back through the doorway.

Over the next half hour or so, the reducing of fractions to their lowest common denominator was made clearer to Mickey than an entire week of classes would accomplish. He had actually grasped what she was showing him within ten minutes due to how easy she made it come into focus, but he was enjoying listening and watching the calm, methodical steps she was outlining for him.

While they worked, he could detect the sounds of someone either cleaning or stacking things in the next room, which he was sure by now was the kitchen. Hearing one particular banging noise, he looked in that direction.

Noticing his interest, Heidi said, "That's Rhett, arranging things in the kitchen."

"Is that your dad?"

A brief pause, then, "Yes. Jed, my grandfather, is in a rear room working on something. He is always tinkering. Maybe you'll meet him later."

The man in the next room suddenly appeared in the same door Blanche had vanished through earlier. He was also thin, but not as trim or as tall as the woman had been. He likewise had a head of hair that didn't look right; a brown bird's nest sitting somewhat sideways on top of a narrow face.

"Hi, there," he said in a friendly manner. "Who do we have here?"

"Mickey Foreman, Rhett. One of my classmates. We're working on math problems."

"That's nice." The man peered curiously at the boy. "Do you live near here, Mickey?"

"Yes sir, Mr. Jones. Just a couple of blocks away."

"Great. I'm sure Heidi and Blanche have told you we're always open to company, so come over anytime." He started to retreat, then added, "And please, you can just call me Rhett."

"Yes, sir . . . um, Rhett."

After he thought the man was out of earshot, his curiosity piqued, Mickey had to know. "Do you always call your parents and grandparent by their first names?"

There was the first hint of indecision he had yet seen in her reply. "Yes, we do," she said slowly.

"We? Do you have brothers or sisters?"

"No," she answered quickly. "I meant *I* do. As Blanche said, we're pretty informal here."

Thinking that was pretty weird, but kinda cool when he gave it a second thought, Mickey decided to get off the subject. It was the only thing that had caused the pleasant girl even a slight bit of discomfort. Plus, he'd lost track of time; his mom might be waiting.

"Well," he said, "I think I've got it now. Thanks for your help, Heidi."

He grabbed up his book, while pushing back from the table.

"You're welcome, Mickey. Any time something stumps you, all you have to do is ask." She looked around the room. "You can stay a while if you like."

"Thanks, but I'd better get home. Mom likes to go out to eat sometimes on Friday nights. She's probably waiting for me right now."

She walked him to the door, and in her typically friendly, but not *overly* friendly manner said, "See you Monday, Mickey."

Walking down the brick walkway back to Louisiana Street, he reflected on his experience of being in Heidi's house, and meeting her somewhat kooky family. Their informality was tricky adjusting to, and the manner of speech they used was definitely different. But it had been an oddly enjoyable visit, unlike anything he had ever encountered. And following the strange treatment from Kristi that had him on a downer, the experience had put him in a buoyant, uplifted mood. He liked Heidi Jones; not in the same way as Kristi, but he really liked her. It would be impossible not to.

CHAPTER EIGHT

Jessica Foreman was indeed waiting for her son to arrive when he finally appeared shortly before six. She was standing in the hallway, dangling her car keys in her hand, giving him the stare he never liked to see; the sign of either anger or disappointment, or both.

"Where have you been, Mick?"she asked, the tone actually showing more relief than impatience. The importance of her two offspring in her life, and their safety, was always on her mind, and always reason to worry.

"Sorry, Mom. I was at Heidi Jones' house. She was helping me with math problems. I guess I stayed longer than I meant to."

"Well, next time, write a note or something, or leave a phone number. You know how I worry." A sigh served as a relief valve of sorts, allowing her to relax some. She reached to tousle his thick hair. "I wanted us to go to Senor Pepe's tonight. You know how busy they are on Friday nights."

"Yes ma'am. I'll do better next time."

"Okay then, we need to be going—and now."

She was leaning into the living room trying to shake up Janice, who was lounging on the sofa. "Let's go, Jan."

The girl, no huge fan of Tex-Mex food, released her own sigh, then reluctantly unwound from the sofa. She launched an accusatory frown in her brother's direction as she joined the other two, as if everything was his fault.

The three members of the compact Foreman family climbed into Jessica's four-year old Toyota Camry, the riders remaining more-or-less silent for the fifteen minute drive north of the downtown area. Dusk was slowly rubbing away the sunlight when Jessica managed to find the last parking slot behind the restaurant. Any later, and she would've had to park on the street. They marched in line like a mother duck and her ducklings to the front of the small town cantina. The building was a modest stucco structure, which had been a western-wear shop in its former life. There were two gnarly wood posts on each side of the front door, with a string of constantly lit Christmas lights strung across the front. People milling around the door was a bad sign; Pepe's was definitely busy.

The popular cantina was owned and operated by Roby Contreras' parents. His father, Arturo Contreras, along with his wife, Margarita, had built a successful business in Cypress Springs after ten years of toiling away at a similar restaurant in Mt. Pleasant. The owner of their former cantina was so confident the Contreras' would do well, he had financed their venture. According to information Mickey had obtained from Roby, which he realized may not be entirely accurate, the loan would be paid off within the next year. But it was obvious the couple was doing very well.

Not surprisingly, there was a waiting list. Margarita Contreras, who served as cashier at the front desk, motioned to Jessica to step to the side of her station to wait, allowing the family of her son's friend to break free of the other customers milling around the small entrance. Mickey managed to catch Roby's eye as he bussed tables, receiving a nod and a shrug from the brown-skinned boy, indicating he could do nothing to speed things up.

Following a fifteen minute wait, during which Jessica served as a referee, keeping her son and daughter separated, stifling their usual restaurant battle, they were shown to a booth. Janice slid in next to her mother on one side, and Mickey commanded the other. Within seconds a huge basket of tortilla chips were placed on the table with a bowl of salsa for each of them. Mickey wasted no time tearing into the stack, dipping with gusto. The females dug in too, but more restrained.

Jessica spent time visiting with customers at nearby tables, some she knew well, and some only vaguely. This was life in small town Texas; especially on Friday nights; and especially for someone in her situation. Working out of town all week, she clung to any acquaintances she had made through her kids' school, or in church, or wherever she could find them. A few she stayed in contact with had actually known Hamp during his younger days. These people were her only connections to the community.

Drink orders were soon taken. Janice and Mickey ordered soft drinks, then watched as Jessica wrestled with her decision. She wanted the cocktail special of the house; a Margarita-Margarita, the name taking advantage of the woman in charge. It would be nice to let the tequila take the edge off the week, but another fact of small town life was that everyone watched, and everyone talked.

"Mom," Janice whispered, "go ahead. You deserve it."

"Thanks, honey." She smiled and spoke to the waiter. "Sweet iced tea will be fine," she said, taking the high road.

It took only a few seconds for the drinks to arrive, since speed was a staple of service in all Tex-Mex restaurants. They soon placed their food orders, then settled back to wait.

Hoping to get some semblance of a conversation started, Jessica zeroed-in on her son. “Mick, tell me about Heidi. And her family. What’re they like?”

He ping-ponged his drink between his hands, sliding the glass over the icy-condensation forming on the table, trying to decide how to describe the Jones’. “Well, Heidi’s nice. She very smart, and easy to talk to.”

“She’s a weirdo,” Janice spoke up.

“No, she’s not. She’s a lot less weird than you. And a lot nicer.”

“Jan! Don’t say things like that,” Jessica admonished. “She’s Mickey’s friend.”

“Well, that doesn’t mean she’s normal.”

“That’s enough, Jan. And Mickey, don’t criticize your sister like that.” She noticed other diners were staring; the debate was getting too loud.

Mickey was shocked that his confidant had suddenly seemed to turn against him. “She started it.”

“I’m aware of that. But I’m finishing it. Right now.”

When the dust had settled slightly, she tried again. “What’re her folks like?”

“I only met her mom and dad. They’re nice too. There’s a grandparent who lives there with them, but I didn’t meet him.”

Jessica considered what he’d told her. “I just figured out which house you went to, the one the Jones’ live in now. Mrs. Frederick’s house, right?”

“Yes, ma’am. First time I’ve set foot inside.”

She thought back over the years. "Come to think of it, I never have either." Time for more prodding. "What do the parents do?"

He frowned. "I don't know. We didn't talk about that. All I got was their names."

"Oh? What were they?"

Taking a swig of soda, he thought back. "Well, the mom's name is Blanche, and the dad is Rhett."

"And the grandparent? You said it was a grandfather, right?"

"Right. I'm trying to remember what Heidi told me his . . . oh yeah, it was Jed."

"Interesting names. Sorta old-school."

Jan couldn't keep quiet. "Sound like clown names to me. Told you they were weird."

Mickey glared at his sister. Why was she acting like this. "Jan . . ."

"I told you to cut it out, Jan." Jessica saw folks looking their way again. "And keep your voices down."

"What do you mean? Old school?" the boy asked his mom, while still glaring at Janice.

"They're names that haven't been used in a while. But they're classy, I think."

"Still say weird," the girl murmured. "Not Mrs. Robinson weird, but weird."

Mickey could only stare daggers at his sister, his mouth hanging open. *How could she*?

Jessica's reaction was instant. "Jan! What are you talking about?"

With a chagrined look that said she knew she had gone too far, and would like to take it back, she replied, "Oh, nothing. Just stuff I heard at school."

"I don't care what you heard. Don't you dare spread gossip about Jane Robinson, or anyone else. Do you hear me?"

"Yes, ma'am. I'm sorry." She shot her brother an apologetic look. He was still too stunned to speak.

In the thankful lull that followed, the food was served, which helped calm things some. Cheese enchiladas for Mickey; beef chimichanga for Jan, and taco salad for Jessica. The food kept their attention away from their arguments, much to the relief of Jessica, as well as the closest diners.

As he took a bite of his first enchilada, Mickey happened to look across the room, just as a couple was being seated in a booth. He nearly dropped his fork when he realized the man was Chief Winston Dunaway. He had only seen the man with his pants down, and from behind, but wearing the same khaki uniform as the other day made it easy to recognize him. He didn't know if the man was married or not, but the woman with him was definitely not Mrs. Robinson.

The chief was speaking to other diners around him. He seemed to enjoy his relative celebrity. Those he chatted with surely knew it didn't hurt to suck up to a police official. For a terrifying second, the man's eyes seemed to focus on Mickey, but then moved on to another admirer.

Knowing his mother might see him staring across, Mickey turned his attention back on his plate. He was already in a funk due to Jan's betrayal. Now he felt this was one of the worst nights of his young life.

Later, as he followed Jan and his mom up to the counter to pay, he chanced one more glance at the lawman, and immediately wished he hadn't. Dunaway was eyeing his mother, giving her an appreciative look up and down as she crossed the restaurant floor. Mickey felt his throat filling with a sickly bile. He looked away.

Making their way to the navy blue Camry, the Foreman family climbed in, all choosing to remain silent during the ride home.

CHAPTER NINE

From Mickey's perspective, the weekend passed without further damage to his psyche, or whatever those things that controlled his inner moods were. In fact, it ended on what he considered to be a positive note.

Saturday was spent in Texarkana, visiting the grandparents on Jessica's side of the family. The cold war between he and Jan persisted during the hour-and-a-half drive to the home of Bill and Louise Knox, situated on the Texas side of the split town. The couple had made their choice years earlier in order to avoid state income taxes they would've faced on the Arkansas side. Thankfully, both of their jobs had kept them on what they considered to be the "right" side of State Line Avenue, or they would've still faced Arkansas taxes. Bill was a now retired postal worker, while Louise continued to work as a school secretary, not wanting to leave a job she really loved.

Mickey always enjoyed the visits with his maternal grandparents; they were nice to Janice and him, and a source of moral support for their mother. The pleasant times spent in the comfortable home were vastly different to the few times they had traveled to Rockwall, Texas, on the eastern outskirts of Dallas, to visit Martha Foreman, the widowed mother of their dad. Henry Foreman, Hamp's father, had died from a massive stroke within a year of his son's perishing in Nam. Martha Foreman tried hard to fill the role of grandmother, but a high-strung temperament, and the confines of a small high-rise condo made the times spent with her torture for both Jessica, her kids, and most of all, Martha. It was sad, but Mickey was discovering that life was like that from time to time, and unfair more often than that.

Sandwiched in between a great home-cooked lunch and a nap-inducing barbeque supper in the Knox's back yard, the three Foremans took in the latest must-see movie. It was *Superman*, which had recently been released, and was right up Mickey's alley. He enjoyed most of it, but wasn't thrilled about the scenes of romance between Superman, a k a Clark Kent, and Lois Lane, which of course, was the best part of the movie according to Jessica and Janice.

When the day was done, Jessica was able to stay awake for the drive home through the gathering Texas darkness, while Janice dozed in the passenger seat, and Mickey stretched out on the rear seat, snoring loudly for an eleven year old.

Jessica was becoming weary of the perpetual battle with her two kids over making the effort to attend Sunday School or church services, and on rare occasions, both. She felt strongly it was worth the struggle to keep them somewhere within range of the straight and narrow. She had been raised in a traditional Baptist church in Texarkana, but had found the Methodist church in downtown Cypress Springs to have a more lenient attitude toward sporadic attendance. Plus, she liked the preacher, who appeared to understand the frustrations in the life of a single, widowed mother, with two kids on the cusp of their teen years. In addition, he, along with many others in the community, recognized the cruel fate of having lost a husband in what a lot of those same citizens believed was a meaningless war.

She would get a break this Sunday morning. Her son and daughter, perhaps both feeling they needed to atone for their Friday night performance, each volunteered to attend without putting up a fight. It was a small victory in the grand scheme of life, but one she gladly accepted. Gratified by their offer, she struck a compromise with them; Sunday school only, followed by lunch at the Stampede, the popular lunch buffet on the highway leading to Mt. Pleasant.

Later that afternoon, with things back to somewhat of an even keel in the household, Janice tapped on Mickey's bedroom door, entering when he said it was okay. Stretched out on his bed, he sat up.

"Look, I want to apologize for what I said Friday night. I guess I wasn't too happy about going to Pepe's that night. Took it out on you. I'm sorry."

He pursed his lips, and nodded. "Okay, well thanks." He mulled for a beat whether to tell her about spotting the chief, and if he should go ahead and reveal the identity of the man he had seen last Thursday. He decided not to. It still might put her at risk right along with him. "No harm was really done. But be ready for Mom to quiz you about what you meant about Mrs. Robinson. You know she's gonna ask, sooner or later."

Jan showed hints of surprise on her face. "Oh, yeah? I'll come up with something that'll satisfy her." She paused, then added, "But I was actually talking about what I said about Heidi, and her family. Shouldn't have said all that."

He swung his feet back and forth. Looking down at his sneakers lying on the floor, he thought randomly, as he had many times, how glad he would be when he finally stopped growing, so he could start wearing boots, like the older guys did. Most parents, and especially his widowed one, couldn't afford to buy western-styled boots for their kids each year until the spurts ended. Somehow, feet seemed to grow faster than the rest of his body. Sneakers at the dollar store were easier to handle each year or so.

"Well, okay. But, Jan, you would really like her if you got to know her. She may look different, but she has no ego whatsoever."

"Maybe so."

"And I've never known anyone who stays so calm, without blowin' up."

"How do you mean?"

He told her of the encounter the girl had experienced with Butch Robinson. "She never flinched."

"Some people are like that. No emotions. And she probably doesn't know about Butch's temper."

He snorted a laugh. "Well, I sure do. Just thought it was a strange reaction."

After they were quiet for a few seconds, he thanked her again for what she had said, and she turned to leave.

"Hey, Jan. What do you know about hair pieces?"

She looked back, frowning. "What? You mean wigs?"

"Yeah. The reason I ask is, both of her parents had patches of hair on their heads that didn't seem to fit right. Just kinda sittin' up there."

She grinned at the image. “Not sure. Some people—mostly adults—like to cover up that they’re losing their real hair. I’d have to see it to answer that.”

“Yeah, I guess. No big deal anyway.”

CHAPTER TEN

Monday, the start of a new school week, a day that would prove to be interesting, and not necessarily in a good way. And at the end of the day, Mickey would be doubting if he understood what had happened, since he hadn't actually *seen* what had happened.

His three morning classes under the new system were: American history, which he enjoyed parts of; music appreciation, which he tried to enjoy since he liked music, but couldn't due to the fact he had no idea how to *read* music; those strange symbols that stood for notes might've well have been Greek (he wondered if Heidi could teach him how to read music—probably not, but it wouldn't shock him if she could); and social studies, which he hated every minute of. But they all had to be endured in order to make it to lunch period.

In the hallway, between music and social studies, Trey caught his attention as they passed each other. "Hey, Mick. We need to talk after lunch."

"What about?"

"You'll see. Let's meet outside again."

Which left Mickey to worry through the entire period about the purpose of the requested meeting. He had a pretty good idea, and it increased his concern. One bright spot during that class was a warm smile from Kristi, who sat three rows over. Maybe things were back to normal, he hoped.

By noon, the weather, which had started out hot and humid, leaving behind the cool tease of the week before, was now darkening; clouds were beginning to form; rain was an almost certainty later. When the gang of four, now including Roby Contreras, gathered on the patio following lunch, this time they occupied a table farther out, since the other was taken. It took only seconds for Mickey to hear his worst fears realized.

"We know who the woman was," Trey said with a sinister grin.

"What woman?" Roby asked, as Mickey's stomach roiled.

Benny did the honors of bringing the Hispanic boy up to speed.

"Wow," he responded. "I've gotta quit taking Friday afternoons off."

That brought a laugh from Benny and Trey, but not from Mickey.

"Okay, Mick," Trey started. "Benny and I checked out the route you had to take home from the D Q. There are three houses that you coulda walked behind when you took that shortcut. The Ludlow place, Butch's house, and the Parmenter place." He paused, watching Mickey's reaction, while enjoying displaying his own cleverness.

"Now the Ludlow house has only one window on the back, and it sits up high on the wall, so you couldn't have seen in there. And old Mrs. Parmenter's got to be in her seventies. Can't see her doing . . . that. So that leaves the Robinson house. It has two windows looking out from the kitchen, which is the room you said you saw her and the man in."

"He's right, Mick," said Benny, his mood not as chipper as Trey's. "It had to be Mrs. Robinson."

Feeling he might lose part of his lunch, Mickey fought back meekly. "You might be wrong, Trey."

"Don't think so, pal. Besides, what's the big deal who it was anyway?"

"The big deal is Butch Robinson. He'll go nuts if he finds out."

"He won't, if none of us ever says anything," Benny pointed out.

Mickey saw some hope. "That's right. We all need to dummy up. If he hears any of us talking about this, he'll kick our butts."

"Wait a minute," Roby said. "You're the one who saw it, Mick. I just found out about it five minutes ago. Why should he come after me?"

"He shouldn't, Roby. But my point is, whoever he hears talking about his mother doing what she did, he'll beat'em silly. Doesn't matter who saw it. And I wish like heck I hadn't seen it."

"Afraid he's right," Benny said. "So, let's all pledge to keep this to ourselves." He surveyed the others.

Mickey was looking straight at Trey Fulghum, trying to read him. If anyone caved, he knew it would be the red headed boy. He then scanned the patio and the doors leading to the lunchroom, looking for Butch. No sign of him, but someone who might pose another problem was exiting the building. It was Heidi Jones, heading their way. She was halfway to the table when he noticed out of the corner of his eye the door at the far end of the building was opening, with Butch Robinson lumbering out. Right inside that door was the boys' restroom, where he must've gone following the usual stuffing of his face. This, Mickey thought to himself, had the makings of a real catastrophe.

"Hi, Mickey," Heidi said as she neared the table.

"Oh, hi, Heidi," he answered as casually as he could.

"Ready for math class?" she asked.

"Yeah, I think so."

"Hope what we covered Friday will help," she said, in a tone as innocent as she always spoke.

The other three boys were now staring at him, smiles breaking out.

"And what exactly did we cover?" Trey asked, the smart aleck trying to look serious.

"Just stuff," Mickey answered, but his real attention was on Butch, now within ten feet. *God, he wished the girl would go back inside.* But she evidently had no intentions of leaving.

And it was too late anyway. Butch approached the table with a menacing scowl directed at the girl. "Thought I told you to stay outta our business."

"Yes you did," she replied evenly.

His face registered surprise for a fleeting second at her response. "Well, you must not've understood. Want me to make you?" He took a step toward her.

Mickey was out of his seat. "Butch, why don't you back off. She's not doing anything wrong."

Butch looked at Mickey as if he was crazy. "What's with you, Foreman? Besides, I ain't gonna hurt her bad. Just mess her up a little."

"Come on, Robinson," Benny finally half stood. "Mick's right. Leave her alone."

"Don't worry about it, Mickey," Heidi said calmly. "You either Benny."

Butch' s face was now puffed up, red with rage. "I'll show you, you bitch!" He reached for her closest arm, but she pulled it out of range with a lightning-quick move, as he stumbled toward her. She then stepped closer and leaned toward Mickey, holding something in her hand.

"Sit down, please, Mickey. You too Benny." She handed a folded piece of paper to Mickey. "Read this later, please." She then placed her hand on the table top.

The four boys seated at the table then stared straight ahead—at nothing—just stared. Five seconds, ten seconds, then fifteen seconds passed. Then they all looked up, swiveling their necks around, looking at each other. They didn't see Butch, but they heard him, groaning and whimpering all at once. They followed the sound, finding the boy lying prone on the concrete surface at one end of the table. Mickey glanced toward the building just in time to see Heidi disappearing inside.

Others sitting at nearby tables looked over, but all they could see was a boy sprawled on the concrete, writhing in pain.

"What happened, Butch?" Roby asked, leaning down to check the boy out.

"Oh, shit. I don't know. Think she burned me somehow. I didn't even see her, or what she did to me."

Mickey looked from Benny to Trey to Roby. "I didn't see anything. Did y'all?"

"No," answered Benny. "I think I just went blank there for a second."

"Me too. But it had to be more than a second," Trey said, "to do whatever she did to him."

Butch was slowly climbing back to his feet. No one helped him, feeling what had happened was his own fault. "I'm gonna have her arrested. Kicked outta school," he said groggily.

The others looked at each other. It was Benny who spoke. "I don't think that's such a good idea, Butch. We were all here, saw what happened. You started it."

"Yeah, but she didn't fight fair. She's not normal. Something's wrong with her."

"Maybe she's not," said Trey. "Normal I mean. So what. And another thing is, do you want everybody to know you got put down by a girl? You'd never live it down."

Butch didn't respond, just shook his over-sized head. He was about to speak when the bell rung, indicating lunch was over.

"Come on, Butch, just forget about the girl, leave her alone," Benny said, heading for the building with the others following.

Mickey realized he was holding something, then recalled Heidi handing him a piece of paper, folded up. It was still in his hand. Stuffing it in a pants pocket, he hurried inside.

Having managed to stay awake during biology, he finally found time to tug the note from his pocket as he settled in to his desk in English. Unfolding it, he read the two lines on the paper, once, then a second time. It read:

> I did not want to hurt him. But I had to make him see
> that there was nothing he could do to hurt me.

Puzzled by the words, he folded it back up, returning it to his pocket. Things had just gotten weird.

CHAPTER ELEVEN

The skies opened up with a vengeance early Monday afternoon, waging a hard rain throughout the area. While the students in Mickey's classrooms watched and listened to the storms outside their rain-streaked windows, Police Chief Winston Dunaway did the same from the security of his office in the downtown city administration building.

The weather he saw outside matched the gloom he felt, a gloom caused by one single person. She had called several times Friday afternoon, and twice this day, at least having the good sense to use the fake name they had agreed to. He hadn't taken her calls, instructing his receptionist to come up with creative excuses. That wasn't going to work much longer; he'd have to talk to the woman soon. He had meant it to be a short term dalliance, as the ones before her had been. There hadn't been that many, but he had been amazed by the effect being a top law enforcement officer had on local citizens—especially women. In fact, Jane Robinson had been one of those woman, who, during the time they had dated briefly in high school, had little use for him.

Rubbing his meaty hands down his khaki pants legs, he thought back to how it had begun, as he had several times lately. He had attended a back-to-school meeting in the assembly hall of Cypress Springs South

Elementary in early August, his part of the program being safety; not taking rides from strangers, walking in groups, speaking up if they saw something suspicious, obeying crossing guards—those kind of things, which he knew were all common sense items. Following his talk, as parents, teachers and students were milling about, she had approached, under the guise of speaking to an old classmate. But long ago friendship wasn't on her mind; it was a new friendship she wanted to kindle, and she didn't bother with subtlety.

It had been fun for a few weeks, and, he had to admit, a real rush. But she was becoming clingy, wanting more and more of his clandestine time. It was getting harder to find the occasion, as well as the place, for their trysts, to the point of lying to his wife and the people in his department. Unmarked cars had to be used, unless they snuck out of town as they had done for two meetings. The cheap motels used in Mt. Pleasant, and Longview to the south, smelled like cigarettes, beer, and a few other odors he'd just as soon not identify. Things were growing stale for him; he needed to end it. He had thought the indignity he had inflicted on her the previous week in her own home would've given her reason to call a halt, but it seemed to have had the opposite effect.

The threat of losing his job, the job he had struggled for eighteen years to achieve, was something very real if she continued to be stubborn. The job of Chief of Police in Cypress Springs, like many smaller towns, was an appointed position. He served at the pleasure of the mayor, who enjoyed a squeaky clean reputation, and expected the same of those he had control over. He wouldn't stand for the sort of scandal the woman seemed to be bent on exposing.

But the greater fear Dunaway felt, the fear that had him constantly looking over his shoulder while with the woman, was her husband, Jim Robinson. He was a large man, with a short fuse. If it came to a confrontation, the relative power and authority of Dunaway's position would not be enough to protect him from the man's rage.

Thinking about that possible outcome made him suddenly rise from his chair. He stood at the window, rubbing a hand across the nape of his neck, beneath the one scrap of hair remaining. Getting to his feet usually helped him to think clearer, but he saw no way—at least no legal way—out of this mess.

The phone console on his wooden desk started buzzing. One of the three lines was flashing. It was his receptionist.

"Chief, that woman, Susan Taylor, is calling again. She says she's not gonna hang up 'less you take her call. Don't know what her problem is, and she won't say."

Dunaway breathed out a heavy sigh, then decided he may as well get it over with. "Give me a few seconds, Shirley, then put her through."

Walking slowly around his desk, the loose buckles on his ankle-high boots flapping, he closed his door for privacy, his frustration causing him to shut it with more force than he intended. Retreating to his chair, he pushed the blinking button.

"Hello."

"Well, glad to see you finally had the guts to pick up," the persistent woman snarled into the phone.

"Jane, you know I can't talk every time you want to call. Got a lot goin' on here. Meetings and what not."

"That's bull, and you know it," she snorted derisively. "And there's no reason you can't pick up the phone and call me once in a while."

Her combative tone let him know this wasn't going to end easily, and probably not today. "How am I to know who's there with you if I call? For all I know, Jim might come back early."

"I told you, he's not due home 'til this weekend. Even longer if his job gets set back with the rain down south. If you'd taken my calls earlier today, I woulda told you he called last night, sayin' that was a good possibility."

He put the receiver against his chest, looking to the ceiling for help. He decided it was time to open the worm can. "Listen, Jane, we gotta be careful. You know good and well, we're gonna be found out. That would be hell."

Another, louder snort. "Hell for you is what you mean. You'd miss that big job you got. No more actin' like a big shot, riding around in that fancy car, suckin' up to the mayor."

My God, what a nightmare. "Let me remind you, this whole thing was your idea, so don't make me out to be the bad guy."

"Hmph! You didn't seem to mind any of it. Especially not last week in my house. You really got into that."

He had to take a couple of breaths, pausing for a couple of beats before responding, although there was no good response to the truth; he *had* enjoyed it.

"Listen, Jane. What I'm trying to say is we need to quit, or at least slow down. Margaret's getting suspicious. And you gotta believe Jim's going to hear about it pretty soon."

"How?"

"I don't know. People see things, and they talk."

"You're gettin' paranoid."

"No. Just being honest about things. Like I told you that afternoon, I swore I heard something, or somebody, runnin' through your back yard. Might've seen us."

Mercifully, another button started flashing. "Hold on. Gotta take this call." She was saying something angrily as he hit the button.

"What is it, Shirley?"

"Robbery, Chief. Out at the Chevron on Highway 13. Guns involved. At least one casualty."

Christ! What else could make his day get worse. "Who's out there?"

"Perkins and Yeager. Two more units and an EMS on the way. There are witnesses, and the subject is on the loose."

"Okay. Tell'em I'm coming out."

As he strapped on his gun belt and stood to leave, he noticed the other light had quit flashing. She had hung up. Best way he knew of to end the conversation for now, but he knew it wasn't over.

CHAPTER TWELVE

Mickey had tried to get Heidi's attention by throwing quick glances in her direction throughout math class, but received no return acknowledgment. He wanted her to explain the second part of her note. He understood the first part, that she didn't want to hurt Butch, or probably anybody else for that matter. It simply wasn't part of her nature. His hope was to catch her after class and quiz her about why Butch couldn't do damage to her.

When the bell rang, he was anxious to hit the hall and hunt her down. By the time he could break free of the milling students, all backed up from the exits since the rains continued to fall, she was nowhere in sight. Elbowing his way to the door, he surveyed the school grounds, but couldn't spot her. She usually rode the same bus home he did, so he made a beeline for the bus, ignoring the rain drops pelting his body. She wasn't anywhere in the bus either. Maybe her parents had picked her up because of the weather. He gave up, knowing he could ask her later. But the meaning of the note was bugging him.

By the time he had made a mad dash from the bus stop near his house in the relentless rain, Mickey was soaked. Glad to find the front door unlocked, which which meant Janice had beat him home, he rushed inside. Dropping his bag on the hall floor, he headed for the hall bathroom, his wet sneakers squeaking on the wood floor. The sound of music wafted through Jan's closed door as he walked by. Today it was Carole King's album, *Tapestry*, which had been big a few years back. Drying his hair with a towel, then removing his wet shirt and hanging it on the shower rod, he grabbed a t-shirt from the chest of drawers in his room. Then stretching out on his narrow bed, he ran what had transpired during the day through his mind.

Whatever had gone on between Heidi and Butch had him mystified, and for some strange reason, a little worried. He knew it had something to do with what she had written on the note. But what? Knowing he would obsess over it until he had an answer, he committed himself to confronting the girl the next day, even if it meant another visit to her house. He felt oddly compelled to return to the place anyway, so that should work.

Then there was the detective work done by Benny and Trey, which he had to admit, was pretty clever. Regardless of the vow made between him and the other three, he didn't think there was any chance the secret would remain within the group, not with Trey included. But there was nothing he could do about that, he didn't think, except to not worry about it. That was a joke; he knew he couldn't control his brain like that.

The D Q was out today. No way was he going to get out in the monsoon again, and he felt certain Kristi wouldn't either. So, with nothing else to occupy his time, he figured boring times called for boring measures, so he went back into the hall, picked up his book bag, and carried it into the kitchen. Spreading out on the kitchen table, he dug into his homework. The sound of horizontal rain striking the kitchen window kept him company.

Later, he heard Carole King grow silent, and Jan soon joined him in the kitchen. Grabbing a can of Dr. Pepper, her favorite soda, from the fridge, she folded herself into a chair across, her limber legs tucked, bare feet on the seat, arms resting on her knees.

"Trouble in paradise?" she asked between swallows. "No Kristi today?"

Mickey wondered if she was back on his side again, or still in attack mode. "Weather's too bad to go out."

"Guess so. Mrs. Wilson pulled up in the carport to let me out, or I woulda been soaked too."

She took another swig, as Mickey drummed his pencil on the table top. It was hard to concentrate with his sister in the room, ready to pounce at any time.

"Anything happen about the Mrs. Robinson deal today?"

Putting his pencil down, he leaned back in his chair. He then told her about the discovery Benny and Trey had made, along with the promise to keep things quiet.

She interrupted. "Wait, you didn't tell them it was her last Friday?"

"Nope. Decided to keep that to myself. Doesn't matter now, though. They figured it out."

"Right. Sorry."

He chose not to describe the confrontation between Heidi and Butch, not wanting her to start in again on Heidi's quirkiness. Plus, he couldn't explain exactly what happened anyway.

"I wouldn't count on all of those guys keeping their word," Jan said. "Benny's okay, but Trey's a loose cannon. Don't know much about Roby."

"Yeah, I know about Trey. Not much I can do about it, though."

Jan eventually returned to her room, and her music, and Mickey made an attempt to get back to his homework. But there were too many other things on his mind.

Jessica finally arrived shortly before six p m. She was loaded down with a bag full of fried chicken and a few other things.

"Mr. Montgomery gave out bonuses for getting that new contract. Used part of mine to stop by Kentucky Fried Chicken."

Within seconds, a bucket was placed on the table, along with mediocre but filling sides and nearly tasteless rolls. The extras didn't matter; it was the chicken they were after. They didn't slow down until there was nothing left but bones and strips of skin the two females had pulled off and left on their plates.

By seven, stuffed and lethargic, they all lolled around the den/living room, watching their new favorite show, *Mork and Mindy*. The idea of a man from outer space in the form of a frenetic comedian who was hilarious was amazing. A perfect show, in Mickey's opinion.

When it ended, Jessie looked in her son's direction. "Hey, Mick. What were the names of Heidi's family again?"

He yawned and replied, "I know the mother's name was Blanche, and the father was . . . oh yeah, Rhett."

"And the grandfather?"

He really had to dig through his memory banks. "It was a short name. Think it was Jeb . . . no, wait. It was Jed, with a "d" on the end."

"Okay. Well, I've been thinking about those names. Those folks must be big fans of old tv shows and movies. Books too."

Jan, who had been only mildly interested in the subject of the Jones', woke up enough to ask, "How do you mean?"

"Starting with the girl, Heidi, that was the name of a girl in the Swiss Alps who was sent to live with her granddad, according to a book written years ago. They made a movie about it too."

Jan nodded. "Yeah, I had to read that book somewhere in second or third grade. Didn't see the movie, though."

"I didn't either," Jessie admitted. "But it was pretty popular."

"What about the others?" Mickey asked.

"Haven't you both heard about one of the main characters in *Gone With the Wind*, Rhett Butler?"

Mickey shook his head, but Jan said, "Yeah, isn't that a classic book."

"Yes. Movie too. About the Civil War."

Jessie seemed to pause for a second, as if trying to decide how to proceed. "Blanche, actually Blanche Dubois, was a character in a famous play by a man named Tennessee Williams. It was *A Streetcar Named Desire*, set in New Orleans, and was about . . . well, you don't need to know what it was about."

"It's not about streetcars?" Mickey asked.

"Afraid not. It's more for adults."

They both looked to her, now wanting more.

"Maybe I need to read that one," Jan said with a sly grin.

"Oh, no you don't. Not yet. You'll get around to it, maybe in high school or college."

"And Jed?" Mickey asked.

"That was the name of the millionaire in *The Beverly Hillbillies*, the man who struck oil in Tennessee—or was it Kentucky?"

"Right," Mickey said. "That's on in the afternoon, re-runs I guess. I like that show."

Jessie had to stop for a moment. Something didn't seem to add up. "You know, I think I must be wrong about that name. Heidi's grandfather had to've been born a long time before that show came out. He couldn't have been named after that old character." She frowned. "Maybe I'm wrong about all of them."

"Whatever," Jan said, standing and stretching. "I've got homework to do before I get too sleepy to do it."

"How about you, Mick?"

"Finished most this afternoon, but got a little left."

She watched the two of them drift from the room. The night had been golden—rare quality time spent with the only two people who mattered in her life, knowing it was a treasure she wanted to mine as long as she could.

CHAPTER THIRTEEN

The speed with which the storms of Monday afternoon and evening passed through should have been a clue that things were about to change weather-wise.

It felt good when Mickey headed to the bus stop Tuesday morning, dancing around and over murky puddles left behind. Not as cool as the previous week's relief, but nice enough to bring promise of a better day ahead.

The unanswered questions about Heidi's take-down of Butch, which he wouldn't have thought possible, and the secret he had unfortunately shared with his posse were shoved to the back of his mind. The day was too nice to stress over things he had no control over.

His classes went well too. Kristi even gave one of her 1000 watt smiles when their paths crossed a couple of times, and the topic of Mrs. Robinson did not come up during lunch, or any other part of the school day. The only thing hanging out there was his inability to find the right moment to quiz Heidi, but he didn't think there was any reason to hurry his interrogation along. But, if he could only somehow predict how the day would come to such an unbelievable close, his buoyant attitude would be tempered in ways hard to calculate.

Receiving one last encouraging grin from Kristi at the end of math class, he enjoyed the bus ride home feeling all was as good as it could be, everything considered. Stopping off at home to stow his book bag on his bed, he walked slowly to the D Q, giving the object of his affections plenty of time to arrive.

Walking briskly into the restaurant, he fully expected to see Kristi Shields sitting and waiting in one of their familiar booths. She wasn't there. He surveyed the other tables and booths; no Kristi. Her sister was there, a sympathetic smile on her face.

Karla yelled over to him. "Have a seat, Mick. I'll be with you in a minute." She was pointing to the booth he had anticipated finding the younger girl occupying.

Sliding into the seat, facing the counter to try and get a read on what was going on, Mickey couldn't deny a sinking feeling in the pit of his stomach. Something must still be wrong, but again, he had no idea. It took a few minutes for Karla to finish up with a customer, then turning to ask a co-worker to cover, she drew two fountain drinks and headed to the booth.

"How you doing, Mick?" she asked, taking the opposite seat while pushing one of the glasses across to him.

"Fine," he replied in a shaky voice. He was more than a little intimidated, due to the girl being five years older, plus being absolutely gorgeous. He had no idea how to talk to someone like her without blowing whatever cool he had.

"Relax" she said, smiling. "Everything's fine. Just wanted to explain what's going on." She leaned toward him, smiling again. "That okay with you."

"Sure." *That was definitely fine with him, since he had no idea.*

"Well, to begin with, you've got nothing to apologize for, since there's no blame. I just wanted to give you some idea of how fragile the ego is of a girl Kristi's age." Seeing a blank look on the boy's face, she added, "Oops, you probably don't know about ego yet. I forget how young you two are."

"Doesn't that mean how you feel about yourself?"

"Close enough. It also means how you think others see you." She took a swallow. "Last Friday, after you left when you couldn't get her to talk to you, I could see she was upset, so I asked her what the problem was."

Mickey nodded, but didn't know what to say, or ask.

"She told me she was afraid you liked somebody else now."

"What? How could she think that?"

"Now, that's where that ego kicks in. She saw another girl smiling at you. And later, she told me you were talking to the girl after school. She likes you so much, Mickey, she's afraid any girl you talk to might replace her."

He immediately knew it had to be Heidi. "That's crazy. Kristi's the prettiest girl I've ever seen. The girl she probably saw was Heidi Jones. She's helping me with math problems. She's a friend—nothing else."

Karla smiled a knowing smile. "I'm sure that's the case. But even if you did like this Heidi girl, Kristi shouldn't let it interfere with you two." She leaned across. "What she and you both have to realize is you're so young, this kind of misunderstanding is going to happen many times over the rest of your lives."

Not knowing how to reply, he simply nodded again, basking in the knowledge of a sixteen year-old, who surely knew all about these things.

"So, I told her not to worry about it. Now, I certainly don't want to see her get hurt, so just be patient with her. She'll get over it, and a couple of years from now, it'll be a thing of the past."

"Does Kristi know you're telling me all this?"

"Yep. She asked me to. She was too shy to talk to you herself." She turned to glance up at the counter, seeing a line forming. "Better go help Bobby Sue. She's about to get slammed. Any questions 'fore I go?"

"Don't think so. And thanks, Karla."

"No problem, Mick. Drink's on me. Like I told Kristi, just hang in there."

Walking in the direction of his home later, Mickey couldn't decide how he felt about what Karla had told him. Part of him was glad that Kristi's bad feelings weren't his fault. Well, they were in a way. He *had* talked to Heidi, but it was an innocent thing. No way should she think it was more than that. He decided he was going to trust what the older sister told him, and hope things would soon get back to normal. Midway home, it suddenly dawned on him; the reason Kristi had been smiling so broadly in class earlier was that she knew the talk from Karla was coming up. Big sister was going to solve her problem for her. He couldn't figure out if he should be resentful over what amounted to a set-up, or not.

So, he now had something else to worry about in addition to one of his friends blabbing. And the image of Police Chief Dunaway lurking was always in his thoughts.

Arriving at Louisiana Street, he again avoided going near the Robinson housc, staying on thc oppositc sidcwalk. Slowing when he drew adjacent to Heidi's house, he was struck by the urge to approach and finally get an answer from Heidi. It couldn't hurt to spend a minute or two solving the puzzle of her note. But the words of Karla came back: "*I certainly don't want to see her get hurt*", meaning Kristi. Would he be betraying the blond-haired beauty if he went in for just a minute? He decided to play it safe, and resumed walking for a few yards. Then he stopped and turned back, his burning curiosity over-riding his better judgment.

He approached the front door, preparing to knock, when the door opened to reveal the pale figure of Heidi Jones, waiting again.

"Mickey, how nice to see you again. Please come in."

Without speaking, he entered the Jones household for the second time.

Heidi closed the door behind him, then lead him to the sofa in the living room. Evidently intending him to take a seat on the aged piece of furniture, she sat at the far end. He didn't think he had any choice but to join her.

Blanche was sitting at the dining room table he and Heidi had used the previous week. Looking down, as if reading something, she turned to acknowledge him. "Hi, Mickey. Good to see you again."

"Hi,' he said, giving a slight wave. "Nice to see you too." He was careful in his reply, not wanting to violate the first name code, which was hard to get used to, completely alien to what he had been taught.

The woman turned back to her reading.

The two sat rather awkwardly for a moment or two, before the girl chose to break the ice. Not appearing to be the least bit impatient, she instead seemed to want to help him get started.

"Was there something on your mind, Mickey?"

He shifted uncomfortably. Grabbing at his jeans pocket, he then realized he'd left the note in his other pair. But he really didn't need it; he'd memorized each word.

"Yeah. Just was wonderin' what you meant in your note, when you said you wanted Butch to know he couldn't hurt you, even as big as he is."

Seated with a view of Blanche behind Heidi, Mickey saw the woman shift slightly in her chair while raising her head a notch.

"Well, that's kinda hard to explain, Mickey," the girl answered slowly. "I simply didn't think he could."

Mickey nodded as if she had given an answer, but of course she hadn't. "I sorta went into a daze after you gave me that note. What happened? How the heck did you put him down?"

Blanche now quickly stood, saying, "Can I get you something to drink, Mickey? Some fruit juice? We have some grapes too."

Caught off guard somewhat by what he saw as an interruption, he managed to say, "No thanks, Blanche. I'm fine."

Before anyone could speak, another person entered the living room through the kitchen door nearest where Mickey sat. It was an older man, who he assumed was the granddad. He was thin, like the others, but not as tall. Whether he had a hair piece on top like the other adult Jones' or not couldn't be determined, since he was wearing a floppy hat, similar to the kind Grandpa Knox wore when he went fishing. But it was obvious the area around and above his ears was slick, lacking hair.

"And who do we have here?" the man asked, grinning at the boy on the sofa.

"Jed," Blanche spoke up, "this is Mickey Foreman, one of Heidi's friends from school, and a neighbor."

"Well, that's nice. Glad to meet you Mickey." The man approached, sticking out his hand to shake.

Grasping the hand loosely, Mickey was struck by how smooth and slick it was; nothing like the rough, gnarly skin of his grandfather, who he judged to be about the same age. He felt a faint, strange tingling as he let go of the man's hand. He nodded and said, "Nice to meet you, sir."

"And polite," Jed said, grinning at Heidi and then Blanche.

"Yes, Mickey is a nice young man. Heidi has been helping him with math problems."

"Well, we all need help with that at times." The man tilted his head slightly sideways, with an inquisitive look showing. "Do you have family, Mickey?"

"Yes, sir. But it's just my mother, and my sister, Janice. She's two years older than me. My dad was killed in the Vietnam War."

"Oh, I wasn't aware of that," said Blanche. "How sad that must be."

"Indeed," echoed Jed. "I'm sure that has been hard on your poor mother."

"Yes, sir."

After a few beats, with no one seeming to know who might carry the conversation, and what it would be, Jed turned to leave. "Don't be a stranger, Mickey," he said over his shoulder. "I'm sure Heidi has told you you're welcome anytime." He then exited, and could be heard going deep into the rear of the house. A door was closed shut somewhere.

With no good reason to stay, Blanche also returned to her seat at the dining room table. It seemed to Mickey, however, she was training one ear in the direction of the sofa, maybe to hear the conversation carried on there.

Mickey decided to tread lightly, thinking he may have somehow touched a nerve in the family. He lowered his voice. "Is there anything you can tell me about what happened after lunch yesterday? I'd sure like to know, if you can."

"Well," Heidi said slowly. "There's really not much to tell. "After I handed you that note, I dodged him when he came at me, and he fell." She added a slight shrug, as if her explanation was all there was to it.

He knew she was not being truthful, and couldn't figure out why. But another thought came to him suddenly, one he should have landed on earlier. "But, Heidi, you must've known you were going to do something before Butch even showed up. You had written the note beforehand, before you came outside. Were you planning to confront him?"

She looked down at her hands, then across the room at Blanche, as if seeking help.

Waiting and hoping for some sort of answer to his questions, Mickey's eyes were pulled away from the girl by the image of a hand reaching onto a counter just inside the kitchen, behind the door at the far end of the room. He looked quickly back at the door to the kitchen nearest him. There he saw Rhett standing next to a corresponding shelf at the opposite end of the kitchen. The distance between the two shelves was at least twenty feet. Knowing the man was alone in that room, Mickey knew his eyes were deceiving him, but again, his curiosity told him he had to check it out. Rising from the sofa, he walked toward the near door.

"Mickey? What is it?" Heidi said, her calmness now escaping her. "Please, don't."

Blanche looked toward the kitchen, shouting words that to Mickey were unintelligible, not actually sounding like words at all, more like gibberish.

He didn't answer—just kept walking. He stepped through the door, just as Rhett was retracting his bare arm into his body. But where there should be skin, there was now a thin, scaly, shiny *tentacle*, the only word Mickey could come up with to describe it. He stood with his mouth hanging open, unable to speak or move.

Rhett looked up, and realizing the boy was there watching, lost his balance for a second, falling sideways. His phony hair fell to the side, exposing the top of his head.

Looking at what he thought would be a bald surface, Mickey instead saw a large hole. Then he looked closer, into the cavity, and saw a green pulsating light, surrounding what looked like—brain!

CHAPTER FOURTEEN

He didn't bother closing the door behind him, running so fast he cleared the two steps on the landing before his feet touched the walkway. Tripping on a brick tilted upwards, he fell to his knees, then quickly rallied to make his escape. He was half panting, half crying as he headed for the street. Darkness wasn't far away.

Nearly to the end of the walkway, he was suddenly aware of two blurs of light flashing past him, one on each side. Something feeling as light as air touched his arm. The blurs came to an abrupt halt, converging in front of him only a yard or so away. The streaks of light instantly transformed into the forms of Heidi and Blanche, standing calmly, yet with looks of concern on their milky faces. His path was blocked; not only by their presence, but also due to the fact he couldn't advance his legs.

"Please, Mickey," Heidi said, "you have nothing to be afraid of. If you will just listen . . ."

He was in full terror mode now, whimpering as he spoke through trembling lips. "Just let me go," he interrupted. "I want to go home!"

Blanche held her hands out, palms facing him. "Mickey, Listen to Heidi. We have no intentions of hurting you, or anyone. And we certainly don't mean to keep you here against your will."

He looked down at his suddenly useless legs. "Then . . . what are you doing to my legs?"

"Okay, we will free your legs. But first, let us explain something." She paused to look behind him toward the house.

Mickey managed to turn his head, following her gaze. He was convinced the others would be coming for him too. Instead, he saw the two men, or whatever they were, standing calmly on the landing, their arms folded.

He looked back at Blanche, who nodded, apparently to Rhett and Jed.

"If you will grant us fifteen minutes to explain what is happening, it will explain what you have seen, and will likely be some of the most thrilling minutes of your life."

His heart was still pounding; chest heaving violently, as if it would explode. He looked up and down the street. Where was everybody? Usually, someone would be on the street, walking, or driving. But not now. Daylight Saving Time was in effect for a couple more weeks, but dusk was falling. Soon, he would be consumed by darkness.

"What are you talkin' about? Explain what?" he asked through snot balls bubbling up.

Heidi spoke up. "Mickey, I am your friend, and you are mine. The only real friend I have in this town. I would trust you with anything, and you should do the same for me. Please, just fifteen minutes, minutes that you will never forget." She smiled the smile he had always enjoyed—until a couple of minutes ago.

It did calm him some. But he still felt he couldn't trust these . . . people. "How do I know you'll keep your word?"

"Okay, we will promise you this," Blanche said. "If you will agree to come inside and listen for those few minutes, Heidi and I will go back in, and your legs will be freed to leave if you change your mind. And I can also promise you that if you do come inside and hear our story, you will not want to leave."

He looked from Blanche to Heidi, and then back once more at the two men at the door. None appeared to offer any threat, so he made a decision to trust them, but with one condition.

"Okay, but I'll need to call home first. They don't know where I am, and they might worry."

"Yes, of course," Blanche agreed. "We will do that the first thing, then we can talk."

He gave a half-hearted nod, then watched as the two of them circled him and began walking toward the house. Heidi brushed a hand softly on his shoulder as she passed by. Immediately he found he could move his legs. So, they had kept the first part of the bargain. Taking one last look up and down Louisiana Street for anyone who might intervene, and seeing no one, he knew he had to keep his part of the deal he had just made.

Shaking his legs to make sure they still worked, he followed them slowly inside.

CHAPTER FIFTEEN

Rhett and Jed were occupying the two wingbacks when Mickey walked cautiously into the living room. Their wigs and floppy hat in place. Nobody objected when he left the door ajar, just in case. Heidi had just re-taken her seat on the sofa, while Blanche stood in the center of the room.

"Do you want to make that phone call first, Mickey?" Blanche asked.

When he nodded, she disappeared into the kitchen, soon returning holding a device of some sort in one of her hands. It was a small rectangular box-like thing, roughly playing card size, appearing to be metal, or maybe shiny plastic.

She handed the boy the box. "I know this is not the sort of communication device you're used to," she said. "but it should work fine."

He stared at the strange gadget for a second, trying to find buttons or dials, or anything he was familiar with. There was nothing visible on the face. "How do I use this?"

"I'm sorry. Just speak the number into that small window at the top." She pointed.

He continued to stare for a moment, then held the device up close to his mouth and yelled out his number. The others emitted something similar to a grin, except for Heidi, who apparently didn't want to make fun of her friend.

"You don't need to talk so loud," she said, as the phone on the other end started ringing.

"Hello." It was Jan's voice.

Stunned by the success of the gadget, he managed to say, "Jan, it's me."

"Hi. Where are you?" Her voice was loud and clear.

"Uh, I'm at Heidi's again. Is Mom home yet?"

"Nope. She just called. Said she'll be late. Big surprise, huh?"

"Right. I should be home in less than half an hour."

After a short pause, his sister asked, "What kinda phone are you on? You sound weird. Like you're standing right here next to me."

Looking at Blanche, he actually grinned. "Yeah, you too. Just a phone of some sort. I'll see you soon." Not knowing how to de-activate the thing he was holding in his hands, he waited for her to say "Bye", and hang up, but heard only silence.

"How do you hang this thing up?" he asked.

"Just punch the middle of the window," Heidi volunteered. He did as directed.

Blanche took the device from him, then retreated to the dining room end, taking one of the chairs and dragging it back to where the others were. She took her seat.

"We'll explain more about our phone later, Mickey," Blanche said. "By the way, we can't accept calls in on it, we can only call out." Seeing his puzzled expression, she added, "We will also explain that later."

Blanche then surveyed the others in the household, getting nods of approval from all three. Except that Jed, shifting forward in his chair, had something to say first.

"I feel the need to say at the outset, Mickey, that I'm not entirely in favor of telling you what you are about to learn. It not only puts us at risk, it also will likely cause you some concerns. Now, Heidi tells us she had complete faith in you, and we hope that trust proves true."

"Jed," Blanche spoke up. "We need to keep in mind that Mickey did not ask for this meeting. In fact we had to plead with him, as I'm sure you heard. It is only due to careless mistakes made by Heidi yesterday, and Rhett only minutes ago, that I feel we must inform him. Otherwise, things could be much worse, as you can imagine."

Jed nodded agreement. "I just thought it should be said."

With that all settled, whatever in the heck it meant, Mickey took his place on the sofa. He did, however, perch on the edge, with his feet flat on the floor, convinced he still might have to make a run for it.

"Well, I think we must begin," Blanche said, "so that Mickey can arrive home in time. As the people in your part of the world like to say, let us 'cut to the chase'." She paused, then smiled strangely at the boy she could plainly see was shaking with apprehension. "There's no easy way to say this, Mickey, but we are not Earthlings. We are from another planet."

The boy wasn't sure at first if he had heard right, but deep inside, he had thought that impossible statement might be true. The force of her words pushed him back onto the sofa. His mouth hung slack. He turned to look at the girl next to him. She was smiling and nodding, as to verify the woman's words.

"Do you have any immediate questions, Mickey?" Blanche asked gently.

About a million, he would realize later. But now, it was all he could do to recall his own name. He looked around the room; they were all smiling, or appearing to, as if this was a common thing to discuss, an every-evening family talk. Finally, he was able to speak, but his voice was even tighter now.

"What . . .what planet did you come from?" He was aware his fingers were digging into his knees.

It was Rhett who spoke. "Well, Mickey, that will be hard to answer. You see, our planet has not yet been discovered by Earth's astronomers, so no name has been assigned to it by your world. If it had been, the name we use to refer to it would not translate to human languages. Suffice it to say, it is a long, long distance from Earth. So distant, it's hard for even us to comprehend." He paused for a second. "Perhaps this will help to describe the physical gap between your planet and ours. Are you familiar with the galaxy your solar system resides in, the Milky Way?"

"Sure."

"The Milky Way is but one galaxy within your universe, which is our universe too. There are many others, but the galaxy nearest to yours is Andromeda, the galaxy our planet resides in, but at its far reaches." Rhett paused again for a moment, appearing to decide on the best way to continue. "Now, when I say the nearest, you must understand, it is still over two-and-a-half billion miles from Earth. It actually contains many more stars than the Milky Way, which means it will probably be centuries before it will be identified and assigned a name by your scientists. However, on a clear night, perhaps with the aid of a telescope, you can actually see parts of Andromeda's spiral swirl from here on Earth. Because of the distances involved, you can imagine how long it took for us to arrive on Earth."

Mickey's mind was still locked in confusion, along with a touch of fear that was beginning to wane slightly. He managed another question. "How . . . how did you get here?"

Rhett turned to Jed, who again shifted in his chair. "Well, Mickey, I will try and answer that without putting more doubts in your head, which I am sure are there now. I was a transport specialist on our planet—at least I was when we departed—tasked with designing and constructing rapid means of transportation. I was, and still am, a pilot of some of those models. Rhett has explained some of the concepts of space-distance, and the same applies to space-speed. Speed is a relative term when it comes to measuring it. On Earth, a speed of a hundred miles-per-hour is considered very fast. To us, it is equivalent to sitting still. In fact, even faster Earth speeds would be nothing to us. Please, understand, I'm telling you this not to impress you, but to expand your thinking a bit." Jed stopped to look at the others. "I think that is enough for now. We have a lot to cover."

That is correct," Blanche said. Then turning to the boy and smiling, said, "Mickey, I can tell you remain fearful, but I need to stress again that we would never harm you. We are not what you would refer to as dangerous aliens."

That received a restrained chuckle from her so-called family, and a faint smile from Mickey. He took a minute to gaze at each of the people—who he now was beginning to see as "beings"—in the room. It was true; there was no look of malice on any of their faces, which he realized now were probably not their actual faces.

"Maybe this is a good time to end this session," Blanche said. "We have a lot more to tell you, but we shouldn't overwhelm you. Will you be willing to return later so that we can explain more?"

"Yes, ma'am . . . I mean, Blanche." He started to ease off the sofa, then decided on one last question. "Those are not your real names, are they? I mean, the ones y'all are using?"

It was Heidi who answered. "No, they are not. We have studied your culture for years, concentrating on your popular media. Each of us has chosen our own names from television shows, movies, and other forms. I enjoyed the story of the young girl who was sent to live in the mountains with her grandfather, much as I have been sent here. We each made our choices based on who we identified with."

Mickey thought about that, turning to look at Blanche. He wondered if she was similar to the woman in New Orleans, the woman his mom was reluctant to talk about.

Blanche, as if sensing his thoughts, actually looked away, as if in embarrassment. He would learn later that was an incorrect assessment.

"My mom figured that part out when I told her your names."

"Your mother must be quite a smart woman," Rhett said. "Very perceptive."

Mickey stood to leave, but just to make sure it was okay, asked,"Can I go home now?"

"Of course, Mickey," Blanche replied. "And please return when you are ready to absorb more of our story. There is a lot to tell you. But, Mickey," she added, looking intently at the boy, "we do ask that you keep what we have told you, and what we will tell you, to yourself. It would be critical to us, and likely you too, if others knew."

"Yeah, I can see it might." He looked first to Heidi, then Blanche and the others, and said, “I promise I won’t tell anyone. They wouldn’t believe me anyway. And I’ll be back the first chance I get.”

After Heidi walked him to the door and said her goodbye’s, Mickey allowed the adrenalin built up in his small body due to what he had just heard to carry him along, half-walking, half-running. Until he reached the corner of Louisiana and Sabine Streets, when the enormity of it all came crashing down around him, and the rush evaporated into the chilled air. The fact that he was the only person, probably in the world, and the only person *maybe ever*, to know what he now knew was too much. His legs buckled; he fell to his knees; he began sobbing due to the sheer weight of what he had heard, and also because of the fear the knowledge brought with it. He ran a hand across his eyes, trying to erase the tears. All those movies and tv shows he had seen of space exploration and inner-galactic battles came roaring back. But those were all fantasies. This, if he could believe the residents in that house, whatever they were, was the real thing.

How could he possibly deal with it?

CHAPTER SIXTEEN

The nearer he got to his house, the slower Mickey walked. He was hoping to erase all evidence of the streaks the tears left on his face. The bottom of his pull-over shirt was damp from wiping. If luck was on his side, his mother would still be late arriving home, giving him time to head straight for the bathroom and wipe the remaining evidence from his face. That was his first priority; the second would be how to handle his rung-out emotions while dealing with his family throughout the night. He would try to keep his distance, but in a small house, that would be tough.

The Toyota's rear end was not visible in the carport, so maybe he had caught a break. Rushing in, he hurried just fast enough to not raise Jan's suspicions on his way to the hall bath, and accomplished his first chore. Then, as he walked down the hall toward his room, he was stopped at his sister's open door when she spoke to him through the sounds of music.

"Hi, shrimp. You gettin' chummy with Heidi?" she asked, with a silly grin.

"What?" He'd forgotten all about the call he'd made. "No. Just hanging out." He wanted to go back at her with a smart reply, but didn't have it in him.

"You're not two-timing sweet little Kristi, are you?"

He immediately recognized a good excuse for any odd behavior he was showing, which he knew would be apparent. "Nah. Kristi was a no-show today, so I went on over to Heidi's."

She looked at him for a moment. "Oh, I see." She then looked closer. "Everything okay, Mick?"

He had never wanted to tell anyone something so badly before, but this was a promise he had to keep above all others. "Nah. Just a little tired, I guess."

Reaching his room, he lay on his bed for a few minutes, thinking back over the last hour, wondering if it had all been a dream. But no. He was wide awake now, and had been then too. Doubts soon crept in. Was it all a hoax? A group of weird magicians playing tricks on him? He barely knew these people. But they weren't people at all; he had seen inside Rhett's head; he had seen his arm stretch across an entire room; he had seen the speed enabling Heidi and Blanche to stop him from leaving; and most of all, he had felt how they could control his body. No, it was real.

As he looked at the ceiling, then out his window, he began to formulate questions he wanted to ask the Jones'. He had to laugh internally to himself; Jan *had* been correct; they were weird, but hopefully in an unbelievably good way.

Climbing off the bed, he wandered into the living room, finding the small book case holding a couple of rows of books. On the second shelf, he found what he was searching for, a book by astronomer Carl Sagan, *The Cosmic Connection*. He had begged Jessica to buy it for him after seeing *Star Wars*, intrigued by the possibilities of space travel and extraterrestrials. Taking it back to his room, he pored through it, looking for anything related to what he had heard earlier in the day. It had actually been less than an hour earlier, but it seemed longer than that. Turning the pages, he recalled how technical some of the book was, finding it then, and now, hard to grasp. But if what he thought he might learn from the family calling themselves "Jones" came to pass, he would soon know more than he could find in any book.

Shortly after six p m, Jessica Foreman finally pulled into the carport. Her two kids could tell right away it had been a bad day. The euphoria of the bonus from the night before was gone now. A tired look and short replies to their questions told Mickey and his sister to give her space. This was a night that called for frozen tv dinners, since she was in no mood to cook.

The table was silent for most of the impromptu meal, which was fine as far as Mickey was concerned. He hated to see his mom out of sorts, but the lack of conversation meant no questions from her about his strangest of all days. They finished up, then went their separate ways: Jessica to stretch out for a nap in the recliner; Jan to her room for more music and homework; and Mickey first to his room, before heading out to the front landing. He needed to be outside. His mind was reeling too rapidly to stay cooped up inside.

The small area outside the front door had two wrought-iron posts supporting the overhang jutting out from the roof. He sat with his back leaning against one of the uncomfortable posts, positioning himself with a view toward the intersection of Sabine and Tyler Streets, which he thought was generally in a northwesterly direction. For some reason he had no justification for, he had always assumed any extraterrestrial activity that might magically appear would be in that quadrant of the heavens. The night was clear, but his view of any stars that may be visiblc were drowned out by the streetlight at that same corner. A telescope would've helped some. He had also begged his mother to buy one when she had agreed to the Sagan book, but she had drawn the line, which he had understood.

After a while, he moved to the backyard, away from the corner light, but any view that may have been there was blocked by age-old trees in the neighbor's yard, their branches bleeding over into the Foremans' back yard. He returned to his post on the front of the house, settling in again.

Soon, Jessica gently closed the front door behind her and came to join him.

"Mind if I sit?"

He straightened his legs to give her room. "Sure. Afraid there's not much to see."

"What exactly are you looking for?" she asked, resting her feet on the first step down, and folding her hands under her knees.

"Oh, just anything I can spot." He thought perhaps an explanation for his sudden interest might be useful. "Some of the guys in school have been talkin' about the universe lately. Wonderin' what it really looks like out there."

"Well, the whole idea's kinda scary, if you ask me."

"How do you mean?"

"I mean, the whole idea of infinity is something I can't come to grips with. Seems like there has to be an end point somewhere." She gave a slight shudder.

"Right. That is hard to figure."

"And, who knows what kind of creatures might be lurking out there."

Mickey simply nodded, finding it hard to suppress a laugh that would be hard to explain. Having *this* conversation with his own mother, on *this* night of all nights, was almost too much to bear. Thankfully, she changed the subject.

"Mick, I've already apologized to Jan, and want to do the same with you. I'm sorry I let my work problems come home with me sometimes. Don't mean to take it out on you two."

"Mom, we both feel you work too hard. No need to apologize to us. I just wish there was more we could do to help you."

She gave her son' shoulder a quick nudge. "Don't worry about that. You and your sister do more than you think." She took a second to peer into the black night herself. "When I think about some of the messes I hear other kids gettin' into, while you two never cause me any problems. Well, I feel blessed."

The gross image of Mrs. Robinson and the police chief jumped into his mind, along with the secret that might cause him trouble. He had to dash it out of his head.

"But, Mom. Most kids I know have two parents. You do more by yourself than theirs do. Jan and I think we're the lucky ones."

Wiping a sudden hint of moisture from the corner of an eye, she wrapped a hug around his thin shoulders. They fell silent for a moment or two, allowing their feelings to mellow out, both staring at the blackness.

"How're you doing, Mick. Don't think to ask you enough. How are you and Kristi gettin' along?"

He took advantage of the question to mention the girl's odd behavior, then gave her a summary of what Karla had explained to him. Then he added, "So, I guess everything's okay. I'll probably find out soon enough."

Jessica digested what Mickey had told her. As valid as Karla's advice sounded to her, she none-the-less didn't think much of a sixteen year old giving pointers about affairs of the heart to her impressionable son.

"Is there anything to Kristi's concern about this Heidi girl?"

"No, not at all. I mean, I like her and all, but it's not the same as with Kristi. Heidi's just a friend." *How dumb that sounded now.*

"Okay then. That settles that."

They sat for a few more minutes, then Jessica drifted inside, with Mickey soon following.

He knew before even climbing into bed later that night, sleep would be next to impossible. There were too many questions about the amazing admission made by the creatures Heidi referred to as "family". What were their real names? What did they actually look like? The sight of what he was sure was a brain inside Rhett's head had him wondering what the rest of the man looked like. And Heidi? Was the shy, pale, friendly girl he had known really covered by scales similar to Rhett's arm? And was she even female? Once you get past all that, why were they on planet Earth? Why in Cypress Springs, of all places?

Tossing and turning, the thoughts went on and on, until the workout of his own brain began to exhaust him. Then, just before he drifted off in the wee hours, a lightning bolt struck him, causing him to sit up in bed, now wide awake again. *The lights in the night skies the previous spring*! That had to be them arriving! The timing fit, based on when Heidi first showed up in school. That would be the first question for them when he made his next visit.

He was now tingling with anticipation. He couldn't wait.

CHAPTER SEVENTEEN

Wednesday morning began with clear skies and a slight breeze drifting across northeast Texas. Of the three members of the tight-knit Foreman family, Janice was the only riser in a decent mood, with no reason not to anticipate a decent day. Her mother, perhaps dreading a continuance of the stress of Tuesday, went numbly through the motions of preparing what she knew was a woefully lacking breakfast; dry cereal, low-fat milk, and buttered toast. She had hoped that by now Jan would have enough of an interest in homemaking to cook every now and then, but that had yet to happen. It couldn't come soon enough.

Mickey was washed out from a lack of sleep, having finally succumbed only an hour before dragging himself from bed. Although his brain was still swirling with questions about the universe, his body was having a hard time keeping up.

They all soon departed in their own usual ways: Jan catching her car pool ride with he best friend's mother; Jessica backing out of the car port for her thirty minute drive to work; and Mickey rushing to the nearby bus stop. Jessica had offered many times to drop him off on her way out of town. He normally declined, saying he enjoyed the chance to catch up with friends and doze every now and then on the way to school. She was pretty sure he refused due to not wanting to be seen being chauffeured by his mother. Not cool.

All-in-all the school day started out fine for Mickey, until meeting up with Benny in the hall between the first two periods. Payne grabbed his elbow, pulling him to the side.

"Hey, Mick. We got problems."

"What?" But he had a good idea. Instantly, he was wide awake.

"Trey spilled the beans."

"How do you know?"

"Couple of kids on my bus were talking about it. Said Trey told them after school while they were walkin' home."

Knowing he shouldn't be at all surprised, Mickey held out hope the damage wasn't total. "Did he tell them how he knew about it?"

"Not sure." Benny peered down the hallway. "Whatta you think Butch is gonna do? I mean, word will get to him soon."

"I don't know, but Fulghum better watch his back." *Me too*, Mickey thought to himself.

Hurrying to music appreciation, the thought kept running through his head about how much his life had changed in less than a week. His supposed girl friend had started acting weird, thinking he was two-timing her. That, he was sure, was not his fault. Just talking to another girl shouldn't put that idea in her head. Karla had agreed with him on that; hadn't she? And accidentally spotting the strange behavior of Mrs. Robinson and the chief; that was an innocent mistake. On the negative side, he *was* to blame for not keeping his own mouth shut and sharing what he'd seen with his friends. That was on him. And now, a family in his neighborhood was telling him they were from another planet! Surely, nothing could top that. He wondered if it was possible for an eleven year old to have a nervous breakdown—the kind he had heard his mom talk about more than a few times. He had no way of knowing things were about to get worse.

Taking his seat in the classroom, pulling out the flute he was supposed to be learning how to play, he noticed Roby Contreras trying to get his attention. The boy mouthed a message. "*Have you heard?*" Mickey nodded, and mouthed back, "*We'll talk at lunch.*"

Later, the other three conspirators managed to corner Trey after lunch, once more on the patio outside the cafeteria. Only this time, they had to stand, huddling on the edge, since all of the tables were taken. During lunch, Mickey could've sworn he heard somebody whispering the name "Mrs. Robinson".

"What the heck, Fulghum?" he asked. "We agreed not to talk about this?"

"Well, I never agreed," the red headed boy shot back. "Didn't nod or shake hands, or anything. This is all your fault, Foreman. You even said that last time, when you admitted you wished you hadn't told us."

"That's pretty lame, Trey," Benny said. "You better not have mentioned Mick's name."

"Of course not," Trey snorted. "Just said I heard it somewhere."

Mickey saw signs of his group starting to fall apart over this. He felt he needed to make sure Trey understood the danger involved. "Trey, if that's true, the word will get to Butch that you're the person everybody heard talking about it. He'll come lookin' for you, askin' questions."

As that sunk in, even the freckles on the boy's face seemed to turn pale. He stayed silent, evidently thinking about the possibility of that happening.

"And," Benny said, "if you rat out Mick to Robinson, you'll have more to worry about than him." He narrowed his eyes, glaring at Fulghum.

Roby, who was the least confrontational of the four, tried to match the intensity of Payne, but he had trouble pulling it off. The best he could do was furrow his dark brow.

With nothing else to be said, and with all four members of the struggling friendship in a funk, the meeting soon broke up. Mickey made his way to his biology class feeling he was not in control of any part of his young life. Maybe he never had been. Going through the motions for the remainder of the school day, he made a decision later in math class he hoped would put things back on somewhat of an even keel. He would meet with Kristi at the D Q if she appeared to be agreeable, and delay by one day a return to Heidi's house. As determined as he was to learn more of their story, the "Jones'" would have to wait. He was exhausted anyway due to the lack of sleep, plus the stress of Trey's betrayal. He needed to calm his life down, at least for one night.

Receiving encouraging smiles from Kristi twice during the day, including toward the end of math, he looked forward to the rendezvous at the usual time and place later. First, he had to explain to Heidi he wouldn't be showing up at her house today. Waiting until he was sure Kristi had cleared the building, he managed to catch up with the milky-skinned girl near the exit. Approaching her, he ran his eyes over the length of her thin body, wondering once more what she really looked like.

His explanation of needing rest was easily accepted, as were most things by the emotionally stoic girl. If anything, she seemed to be expecting the postponement.

"That's fine, Mickey. Whatever suits you." Leveling a knowing grin, she added, "Please don't let me or what we want to tell you interfere with you and Kristi. I know the time you spend with her is very important to you."

He was pretty sure he had a stunned expression on his face, but quickly realized he shouldn't be surprised at anything the girl knew. She had her ways.

"How did you know . . .?"

"Maybe the next time you come by, I can cover that also. We are anxious to tell you more, and I am sure you have many questions, so just come by anytime." With that, she smiled again and turned to the exit door.

Staring as she walked away, he now wished he had kept to his original plan of visiting the girl's house that afternoon. But he was tired, and knowing he had to regain his standing with Kristi, he went in search of his bus.

It was hard to tell who felt more uncomfortable when Mickey slid into the booth at the D Q. Kristi's blushes and nervous manner showed embarrassment, indicating she understood she had mis-read Mickey's attentions toward Heidi. Mickey, knowing he had to tread lightly with the girl's fragile ego, as explained to him by Karla, he did his best to reinforce her trust in him. He sipped at the soft drink she had waiting for him when he arrived.

"Heidi is just a friend, Kristi," he pointed out. "I'm sorry if you thought anything else." Even as he said the words, he realized he had a choice to make; one that Karla's advice hadn't covered. If he was honest and told her he would be seeing a lot more of Heidi, at least in the immediate future, she would surely want to know why, which, of course he couldn't answer truthfully. But if he dummied-up about it, and told her nothing, would that be a betrayal of the relationship? Probably so, but that seemed the easiest way to go. He didn't see how he had a choice. Crap. Life could really suck, he was discovering.

So after their normal gossip about school work and classmates, with his nerves teetering on the edge, they said their goodbyes half an hour later. Mickey waved at Karla as he left. She waved back, giving him a thumbs-up. Even that gave him a flash of guilt.

CHAPTER EIGHTEEN

Quit worrying about your damn image. No one's gonna find out, unless you screw up and let something slip."

During a weak moment, Chief Winston Dunaway had allowed Shirley to patch the call through from "Susan Taylor". It was barely nine a. m. Thursday morning, and his day was already being ruined. But if he put her off, she would just keep trying. She sounded more insistent than usual, and more desperate, which was cause for concern. Desperate people do desperate things, as he had learned over a two decade career in law enforcement. Brain-dead drug users, people who had lost jobs, gamblers in over their heads, and even jilted spouses and lovers—he had seen it all. And that was his situation, a soon-to-be jilted lover. If he couldn't warn her off, at least he might be able to appease her.

"You don't need to fret about me messin' up," he said as calmly as he could. "Like I told you, I've got too much to lose. You need to get a grip."

Jane Robinson breathed fire into the phone. "I'd like to get a grip of your neck, sayin' things like that."

"Okay, okay. Let's settle down a little. I only meant to say I don't make mistakes." *Except gettin' mixed up with you,* is what he wanted to say. "Just tell me what's on your mind."

His tone seemed to have some affect. She spoke more evenly. “What’s on my mind is that Jim’s definitely coming home tomorrow night. He told me he’ll likely be here for a couple of weeks, unless they call him back. If you really want to slack off some, we need to see each other once more before he shows up. Won’t be so easy after that. Maybe this afternoon, or sometime tomorrow.”

He placed the receiver against his chest, trying to decide if a couple of weeks of relative peace was worth going through the hassle of planning something on such short notice. And he well knew, her hint at two weeks of not meeting couldn’t be trusted. After all, they had met several times early on with Jim in town.

“Well, if you think it’s a good idea, let me see what my schedule looks like.”

“You mean you don’t already know? You’re the Chief. Don’t you make your own schedule?” She was really getting mean. More desperation was showing.

“Of course I do, but there are always changes.”

He spoke again quickly, before she could spew more venom. “Let me think a minute.” Waiting an appropriate length of time, he said, “Here’s what we can do. I’ll go alone to lunch at Wendy’s, on Marshall Street. About twelve-thirty, you circle through the lot there. I’ll be standing out near my car. Pull as close to me as you can, and I’ll climb in your car. We’ll go to your place. Should give us a couple of hours before your son gets home. Then you can take me back to my vehicle when we’re done.”

“When we’re done? You make it sound like a chore you gotta take care of.”

He had to hold his tongue for a couple of beats. “That’s the best I can do. You gonna be there?”

A long, tortured sigh, then, “Alright.”

CHAPTER NINETEEN

Trey Fulghum avoided eye contact with Mickey whenever they encountered each other during all of the Thursday school day. But whether or not that meant he had shared the identity of the observer of Mrs. Robinson's strange behavior with anyone, Mickey was sure it was only a matter of time, just as he had known Trey would be the first to crack about the story itself. Maybe it was his imagination playing tricks on him, but there had been several times between classes in the halls, and during lunch period, he could again swear there were whispers of either the woman's name, or even worse, her son's.

It was only the anticipation of answers to questions he might receive from Heidi and her alien family later that afternoon that allowed him to put those other concerns to the side, but only briefly. He planned to cut his daily visit with Kristi short, hopefully without hurting her tender feelings, and spend as much time as possible in the house on Louisiana Street. He had wanted to write down the things he had in mind to ask, but decided it was too risky; prying eyes might wonder about the odd nature of his questions. They were all cemented in his head. When the final bell rang at 3:30, he wasted no time finding his way to his bus. He didn't bother giving a clandestine message to Heidi that he would see her later, trusting her repeated promise that anytime was fine. When she climbed aboard the yellow vehicle after him, passing by him as she took a seat in the rear, a quick grin was exchanged between them. The route home was at the same speed as always, but seemed an eternity to Mickey.

A short stop at his home to deposit his book bag on his bed, then it was on to the Dairy Queen. He and Kristi were more relaxed this time; things seemed to be back to normal. Their usual idle, casual conversation was moving along fine, until the subject matter suddenly changed.

"But I did hear something strange today, Mickey," she said innocently, tightening her blue eyes as if confused. "I don't really understand what it was all about, but a couple of the girls in home-ec were talkin' about a man and a woman doing something awful. Just heard bits and pieces about it, but it sounded gross."

Mickey's legs instantly grew tense, the stress of what the "awful" thing was radiating downward. He hoped his face wasn't betraying his panic.

"Wow. What were they doing?"

She shook her blond locks. "I'd rather not say what I heard. Like I said, it didn't make sense to me. I don't know about that kind of stuff."

"Did they say who they were? The man and woman, I mean."

"No, but it sounded to me like one of them had a child in Southside. But no names. Have you heard anything like that?"

"Nope," he answered, probably too hurriedly. "But, gosh. If you hear anymore, let me know. I'll ask around."

"I don't wanna hear anymore," she said, with a sour expression. "Wish I hadn't learned what I did."

You and me both, he thought. "Well, just try and forget you heard it."

One of the Jones' was missing from the living room when Mickey settled in on the sofa once again just before 4:30. Jed was somewhere in a back room again, tinkering with something. Heidi was in her seat on the sofa, with Blanche and Rhett occupying the wing back chairs, which had been pulled closer to the couch.

"Mickey, do you have any questions to start us off today?" Blanche asked, evidently leading this session.

He looked at the three of them, still in awe of who they claimed to be. "Yes, I do. About seven months ago, there were bright lights streaking across our skies at night. From north to south, it seemed. No one had any idea what they were, except maybe meteorites. But no marks were found where they might've hit land." He paused before going on, anticipating the answer. "That was around the time I first saw Heidi in school, at the end of the school year. Was that how y'all got here?"

She smiled and nodded. "Yes, Mickey. That was us arriving in your area."

The truth of that affirmation nearly took his breath away. Inhaling and then letting air out, he followed up. "But the lights were up there in the skies for two nights. Did it take two trips?"

As Rhett chuckled softly, Blanche answered, "No, just one trip to actually make contact, which was the second night. The first night was to get a picture of the surroundings, to insure it would be a compatible place for us to reside. You could refer to it as a flyover."

Mickey nodded, thinking back to those nights. "But nobody ever found any burn marks or wreckage, not even any broken tree limbs anywhere."

Rhett chuckled again as he answered. "We are able to use what you might call a soft landing. Jed is an excellent pilot. No damage is done to the local environment. That is important to us."

Thrilled at receiving the answer to his first inquiry, Mickey was having a hard time recalling the others on his mental list. One soon came to mind.

"Did you come here straight from your home planet, or somewhere else?"

"Somewhere else would be correct," Rhett said. "We have relocated here from one of your northern states, Michigan. Actually an area referred to as the Upper Peninsula."

That was information Mickey hadn't expected. Another planet had been the reply he thought he would get. It adjusted his view of things, but he couldn't figure out exactly how.

"Why did you choose Cypress Springs? It's not a very exciting place."

"From our first trip across your landscape, this area seemed to be rural, with plenty of vegetation in the form of trees and smaller varieties of underbrush. Those are two things we find necessary. We choose to keep a low profile that might not be possible in larger cities. We found the same conditions in Michigan."

The boy thought about that. "Is that because you don't want to stand out?"

Rhett nodded slightly. "Primarily, yes. Another factor is that the thick woods offer useful locations to conceal our method of transport."

Blanche shot a quick, questioning glance at Rhett, which he responded to. "I think we can move ahead into areas we were not going to cover until later. Mickey seems to be worthy of our trust."

Looking from Blanche to Rhett, Mickey quickly agreed. "You can definitely trust me. I want to learn everything you can tell me." A new question came to mind. "Why did you leave Michigan?"

"It was simply time to relocate," Blanche said. "We are here to observe different cultures and environments. But to be honest, the winters that far north were not kind to us. Our planet's climate can be controlled to keep our temperatures moderate, and we found ourselves reacting poorly to the cold."

"So," said Mickey, going back to Rhett's statement, "your space ship is hidden out in the woods somewhere?"

"Yes," Rhett said, smiling. "But it is hardly a space ship. You will be surprised perhaps at how small it is."

It sounded as if Rhett was saying he would be shown the craft, whatever it was. That was almost too much to comprehend. He calmed himself with the question he originally wanted to ask Heidi.

"Heidi, you never finished telling me what you did to Butch, or at least I don't think you did. He didn't just trip, did he?"

"No, and I apologize for not being entirely truthful. We do not normally use evasive answers. At that time, we had not decided to reveal ourselves to you, so for our protection and yours, I did not give a full answer."

He nodded, waiting for her to continue.

"We are born with an inner characteristic, or perhaps mechanism is a better word, which produces an electro-magnetic energy that protects us, but only when we call it up by wishing it so. This force, by the way, is in existence here on your planet. It is utilized in most machinery you humans have developed." She glanced at Blanche, and then Rhett, apparently seeking approval to continue. They both nodded.

"This energy can be at differing levels, to produce the effect we think is necessary. In the case of your friend, Butch, I only wanted him to feel a small amount of energy, so I used very little. We can also use materials, such as metals, to conduct this power, which I did by touching the table surface. That is why you and your other friends that were seated were immobilized until I had done what I had intended to Butch. As I said in my note, neither I or the others want to hurt anyone permanently, but sometimes, a lesson needs to be taught. But only for our own protection."

Mickey was aware his mouth was hanging open in amazement. Clamping it shut, he asked, "So you were born with that ability?"

"Yes. Well, we are not actually born in the way you humans are, but are created in a similar way. That capability is embedded within our inner being. We don't have the characteristics you are instilled with, such as emotions, like fear, anger, love, hate, jealousy, and envy. Those are feelings you are able to call on when needed, whether voluntary or not, none of which we possess."

Her explanation answered a lot. Her calmness under Butch's verbal attacks was now explained. "Is that the same energy y'all used the other night to keep me from running away?"

"Yes, Mickey, it is," Blanche said, leaning toward him in her chair. "We certainly hope you were not in pain because of that. We would have allowed you to go if you had continued to insist, but we understood how exciting what we would tell you would be. And, to be truthful, we were afraid you might give us away before we had a chance to explain."

Her comments brought another question to mind. "Why do you feel you have to stay hidden? Are you afraid of us? Afraid of humans?"

"As Heidi explained," Blanche said, "fear is something we do not possess. But it is the fear humans might have of us, if we are exposed." She paused for a second to let that sink in. "Are you familiar with movies, such as *Close Encounters of the Third Kind*, and one made many Earth years ago titled, *The Day the Earth Stood Still*"?

"*Close Encounters* I know. I saw it, but never heard of the other one."

"Well, we believe the combative stance Earthlings displayed toward the terrestrial visitors in those movies, plus many others, represent the attitude we might face, which is actually fear of the unknown. We realize those are fictional works, but we do not wish to inflame those doubts about our presence, if they do exist."

"What would *your* planet think about people from other worlds arriving on yours? Would it be different?"

"Yes," spoke up Rhett. "It *has* been different. Natives of two other planets in our galaxy have visited us. We now have cooperative relationships with them. We do not fear or doubt each other."

That brought a quick follow up from the boy. "I have always wondered if there was life on other planets in the Milky Way. Scientists are always looking, and hoping. Do you know?"

"Unfortunately," Rhett said, "during our trip through your galaxy, including your solar system, we saw no evidence of life as you would recognize it. I would like to say otherwise, but it appears Earth stands alone in that category."

Mickey was sure his own disappointment in that opinion would be nothing compared to Earth's astronomers. It had always been a hope hanging out there.

Blanche interrupted his thoughts.

"I think that should be enough for this session. We have given Mickey a lot to digest, and it will take some time to do so."

"Okay," he said. "But I want to hear more. A lot more. You and Heidi were right, Blanche. This is the most exciting thing that's ever happened to me."

"That is good to hear, Mickey. We want to share with you what our lives are like, not only on Earth, but on our planet too. So, keep gathering questions for us, and we will continue next time."

CHAPTER TWENTY

During the days that followed, a comfort level began to develop during the sessions of discovery held between Mickey and the occupants of 710 Louisiana Street. In truth, the comfort established was all on the part of the wide-eyed boy, trusting the four members of Heidi's group with their tales of worlds that he, up to now, had only hoped existed. The gentle, and so far friendly aliens, didn't have to worry about comfort levels, since they never felt anything approaching discomfort—that would involve something similar to feelings, which, according to them, they did not possess.

Getting through each school day was the hardest part for Mickey. It was a constant juggling match involving Heidi and Kristi. He had to be sure any knowing grins and nods of his head between he and Heidi were given out of Kristi's vision. He was still maintaining his almost daily reconnoiter with the blond-haired girl, and didn't want to insult her with any signs she might find suspicious involving the awkward new-comer, who was busy introducing him to wonders no one else knew even existed. He managed to keep it together, anticipating the next meeting with the family he was now convinced were aliens.

At five p. m., Friday, when he arrived at their home, he had a minute alone with the girl, as they waited for the others to join them. He took advantage, asking her a question he knew he may not want to hear the answer to.

“That first night, when you and Blanche asked me to stay and listen, what would’ve happened if I’d said no, and left. She told me I could go if I wanted to, but it couldn’t have been that simple.”

There was only a slight hesitation. “We had discussed that possibility before that occurrence. To be clear, I always thought we would reveal ourselves to you eventually. Our opinion was that if we thought you could be trusted to not discuss any suspicions you might have about us, we would allow you to leave. Of course, we would have to monitor your actions in the days after to see if our trust was well founded.”

“Monitor? How?”

“By observing how you interacted with your friends, and family, if possible. It would give us an idea if our second option was necessary.”

Mickey felt the first tinge of danger since that first night, when he had decided to stay. What might they do? The mention of his family bothered him. “What was the second option?”

“We would have no choice but to relocate once again, for the reasons we discussed in our previous meeting.” The words were spoken without even the hint of remorse or potential regret at having to leave, but were words which struck Mickey as depressing as anything had since the loss of his dad; possibly more so, since he had been so young at the time.

“You would leave? Where would you go?”

"We have several choices to consider, but none of which I will share with you in case the time ever comes."

He leaned back on the now-familiar sofa, absorbing what she had told him. "In that case, I'm sure glad I didn't run away."

"We are too, Mickey," said Blanche as she came into the living room, followed by Rhett and Jed. She had evidently been close enough to overhear the conversation; perhaps wanting to hear it. "We are enjoying the opportunity to share our story with you."

Rhett and Jed took their seats across, and Blanche again pulled a chair from the dining area to sit near Heidi and Mickey.

It was Rhett who spoke next. "Now that you have that question answered, which was a good one, by the way, what else is on your mind?"

He leaned forward. "I do have a couple." Actually, three, but the third might be pushing things too far. "Can you tell me what you look like? I mean, without your, er . . . human skin on."

The four inhabitants of the house looked at each other, apparently caught somewhat off guard. It was Jed who finally volunteered to speak. "That might be too much for you to experience at this time, Mickey. But I do have one thought. Let me think it through, while you pose your other questions. I will consider it and give you an answer before you leave tonight. Does that sound fair?"

He nodded his response, glad the request wasn't completely canned. It dawned on him that Jed, being the elder, might be the decision maker of the group. He then asked about something that had stayed with him from the first day of the reveal.

"You mentioned before how far your planet is from Earth, and it sounded like it would take a while to arrive on Earth. How long did it take?"

"Before I answer that, recall that we explained that time and speed are relative terms, meaning that ours are different than yours, by a large factor."

Confused by the words, "relative terms" and "factor", Mickey tried to give the impression he was following along. "Okay."

"Well, in Earth years, our voyage lasted almost fifteen years, nearly a decade-and-a-half. If your scientists were to launch a craft in search of our planet, the time in space would be many times longer. In fact, if it were a manned vehicle, no one would survive to witness the arrival."

That fact was somewhat shattering to Mickey, who had always dreamed of space travel. What Jed had just described was mind-boggling, and again, depressing. It also made him think of something else. He turned to look at Heidi, who seemed to anticipate his next question.

"That means . . . how old are you, Heidi?"

"I will only say that I am much older than I appear, Mickey. However, in my current form, I feel the same age as you." She smiled her usual gentle smile. "I know that is hard for you to imagine, and I wish you would not think about it too much. I want us to remain friends who share classes."

"I want that too, Heidi. It just seems, well, weird, thinking about it."

The room fell silent for a moment, with Mickey adjusting to what he had heard, and the others observing his reaction.

Something else suddenly occurred to him. "But you said before your vehicle you arrived in was small. How could you travel such distances and over so many years like that?"

"Another excellent question, Mickey,"Jed said. "The capsule-like device we arrived in here is only our module for moving around your planet. We had a much larger ship for our long journey. It contained all we needed to sustain us for the duration. When we reached the outer limits of Earth's atmosphere, which your astronomers refer to as the exosphere, we were jettisoned into orbit in our smaller craft until we chose to descend. Our mother ship then returned to our world."

Mickey's brain was spinning so rapidly at all he had heard, he had no idea what to say or ask next. He simply shook his head slowly, which actually produced something resembling smiles around the room.

"I think we have hit upon another stopping point," Blanche said. "I believe Mickey requires time to digest what we have covered tonight."

Mickey was prepared to agree, and was starting to rise, when Jed spoke up.

"Mickey, do you still wish to see us in our natural state?"

"Yes. If that's okay with you." He returned to his seat with new excitement.

"Here is what I propose. The three of us," he said, motioning to Rhett and Blanche, "will go into the kitchen area. Only one of us—you will not know which—will emerge as you have requested. Heidi will remain with you so that you will know it is not her. I believe it is important you continue to see her as you do now."

Mickey nodded, and gave a meek “Okay,” now uncertain if he was really ready for what he might see.

The three designated aliens rose and disappeared into the nearest opening leading to the kitchen. Mickey looked at Heidi, wishing he could hold her hand for support, but knowing it would not be the same as holding a human’s hand. He waited, hardly able to breathe.

No more than thirty seconds had elapsed when a form appeared in the doorway they had vanished through. It then entered the room so that he had a total view. To the boy’s eyes, it resembled a fuzzy glow of greenish light, sparkling on its outer edges. Within the form was the outline of a body which could almost be identified as human, if it were not constantly evolving into alternate shapes. He could make out what had to be eyes, but there were four of them, two above, and two below, on an elongated face. They were all trained on him, and for a mesmerizing second, he expected them to launch some sort of rays, zapping him into oblivion. Then they blinked, morphing into what appeared to be crescents, as a human’s eyes would if giving a friendly smile. It reminded him of how his mom’s eyes crinkled when she smiled. As he was able to focus on the shape, he now saw what he had seen inside Rhett’s head that night; a mass resembling pictures he had seen in text books of a brain. Only this one seemed to be floating within the green lights.

The form suddenly began spinning, slowly at first, then increasing in speed, until it looked like a spinning top. The glow from the sparkling edges seemed to light up the room. Trying to follow it with his own eyes, Mickey realized he was laughing hysterically at the sight. A quick sideways glance at Heidi saw her smiling in approval.

The spinning came to an abrupt halt. An extended limb of sorts, appearing to be an arm, similar to the arm of Rhett's on that first late afternoon, reached out as if waving to him. He thought it might be Rhett's, but figured they all probably had the same type. It was quickly retracted, and the shape exited the room so quickly, there wasn't even time to catch the movement. Mickey had witnessed that speed before; Heidi had used it, on Butch.

He leaned back against the rear of the sofa. He could only shake his head once more, uttering, "Unbelievable!" He was now near tears, as he had been that first day of their reveal to him, overcome by the unworldly display.

"Are you okay, Mickey," Heidi asked. "That last part was just for your entertainment. Showing off, is how you would label it."

He looked at the girlish figure sitting next to him, wanting to ask if she was like that. Then agreeing with Jed's earlier words, he really didn't want to know.

The others were back in the room within another minute, taking their seats, hair piece and floppy hats back in place.

"Well, Mickey," Blanche said. "I hope that did not alarm or frighten you."

"No. It was amazing."

"It was nice for us too, having a chance for one of us, at least, to free ourselves of our human form. They can be rather bothersome."

"I can see that now."

Blanche peered around the room, as if surveying the others. They all gave brief nods. "Mickey, we have a proposal of sorts for you."

Unsure of what was coming, he again simply nodded.

"Years from now, when you are grown to adulthood, or perhaps even before that, you will surely want to share what we have shown you, and told you. We have asked you to keep our existence confidential for now, but we cannot expect you to do so indefinitely. It would be beneficial for you to have someone, someone you would trust to also keep our secrets, to experience our story along with you. They could serve to validate what you have discovered. We feel this is too large a burden for you to carry alone." She paused for a few seconds. "Do you understand what we are offering?"

"You want me to bring somebody else to listen in with me?" The thought of that gave him a feeling of encouragement, but also posed a problem. Who could he trust enough for that. Who could they trust? Heidi interrupted his thoughts.

"How about Janice?"

Janice? Surprisingly, he had to think for a moment who she was referring to. "Jan? My sister?"

"I have only seen her from a distance, but she seems to be intelligent, and perhaps open-minded enough to accept what we have to say. And other than your mother, likely the person you are closest to, and that you can share our secrets with."

Mulling the possibility of having his sister know what he knew, and considering how she had waffled so much lately, he wondered. But if not her, then who? He then thought that it made sense to keep the secrets of their amazing neighbors in-house, so to speak. He could certainly control her better than his friends, who he only saw part of the time. Of course, controlling Jan wasn't at all easy. But it did sound good to know he wouldn't be shouldering the burden of his knowledge alone, so he made the decision.

"Okay. But let me think about it over the weekend. I was going to mention anyway that I need to spend time with my family and maybe my friends the next couple of days. Mom might be gettin' suspicious of all the time I'm here."

"Certainly, Mickey," Jed answered. "That makes sense to us. Enjoy your weekend, and visit us again whenever you are ready. But, remember, besides whoever you choose, we ask you to please guard our secret from others."

"Don't worry about that."

CHAPTER TWENTY-ONE

Saturday morning started with a neighborhood football game, the neighborhood being expanded by several blocks. Benny Payne called shortly after nine a. m., telling Mickey a group of seven kids, Benny being one of the seven, were going to meet at the vacant lot they used often for pickup games. The lot was roughly halfway between Foreman's and Payne's houses, which situated it about four blocks from both. A promise that Butch Robinson would not be included clinched Mickey's agreement to show up to make an even eight.

Four men per team was not ideal, but enough to get battered around some, show off throwing arms, and if lucky, go home with bloody jerseys or bruises that would soon turn purple. Battle scars were good signs, especially if they lasted until Monday to be displayed at school. The teams were fairly even based on sizes and abilities, except for two members of the side opposing Mickey, Benny and two other classmates. Trey was on the other team, which was just as well with Mickey . . . he was still pee-ode at the guy, as were two black kids who lived nearer downtown. One was the only true athlete on the field, hard to catch, and elusive to bring down when he was caught.

By eleven, with his bowl of corn flakes wearing off and hunger pangs hitting, Mickey trudged home, with no more to show for the morning than muddy shoes and mud and grass-stained sweat pants. As he walked, he used the time to consider his current dilemma; that being whether to recruit Jan to join him in his adventure or not.

To begin with, she had been acting so squirrelly lately, he couldn't decide if he could trust her or not. But in the usual case, she was the most trustworthy person he knew, except for his mom, who in this case was not a candidate. There was also the chance she wouldn't want to be involved once he told her Heidi was part of the deal. It would take a real sales job to convince her, and the only way to do that was to tell her what he had learned so far without sounding half looney. And if he failed, there was a real danger she would go straight to their mother, claiming her brother had indeed gone nuts. In the end, Blanche's sensible words that he needed to share this discovery with someone made perfect sense, not only now, but in future years. Otherwise, he might *indeed* go insane due to the burden of keeping it to himself. So by the time he arrived home, he knew he had to talk to her.

But *when* to tell her was the trick; it had to be when they were alone, out of their mother's earshot. This was in the category of things he normally found easy to procrastinate over, letting it hang out there through the weekend. He decided to look for a time this afternoon, attack it like yanking a band aid right off, and bite the bullet that could very well explode in his face.

His chance came mid-afternoon. Jessica Foreman announced she was going to use what was left of her bonus from the week before to do some shopping. When the Toyota had cleared the driveway, Mickey began pacing in the hallway, crossing back and forth in front of Jan's door, which was slightly ajar. She was on her bed, on the phone Jessica had recently installed for her, talking to one of her gossipy friends. Knowing his window of opportunity might be limited, he finally managed to catch her annoyed eye. Holding the receiver away, she mouthed, "What do you want?" through the open doorway.

"We need to talk," he mouthed back.

Giving him a sour look, Jan told her friend, "Gotta go. My nerdy brother wants to tell me something." Hearing the other girl's reply, she made another face, then hung the phone up.

"Barbara Sue said to tell you she thinks you're cute. Says a lot about her judgment." She rolled her tongue out for emphasis. "Now, what is it?"

"Okay, well, then tell her thanks," he said absently, entering the room and leaving the door open a little wider in order to hear better if their mom returned early. He removed the record player from her desk chair, pulling it close to her bed to sit.

"There's something I gotta tell you," he said, lowering his voice just in case. "And you gotta believe me, Jan, this is the most important thing I've ever told you, or anybody."

"Is it about Mrs. Robinson?" She leaned in at the prospect.

"No. Much bigger than that."

"What could be bigger than that?" she asked incredulously.

Well, here goes, he thought to himself. "It's about Heidi, and her family."

"Oh, that again. I already apologized about what I said. Mick, you don't need to . . ."

He waved his hands to stop her. "No, no, not that. But you were right about them being strange. Only they're a lot stranger than you thought."

She furrowed her brow, showing a worried look. "Stranger? How?"

"Alright, but you need to understand I'm not making anything up, okay?"

She paused and leaned back a little. "I'll try."

He shook his head rapidly. "You gotta do more than try, Jan, because what I tell you is gonna change everything, and I can't take it back once I say it."

"What are you talking about?" Genuine worry was showing now.

He took a deep breath, then dove right in. "The truth is, Heidi and her family are aliens." He sat back to watch her reaction.

"What? You mean, like from Mexico?"

Confused for a second, he then explained, "No, from somewhere else." He thought that might do it. But it hadn't.

"From where then?"

He pointed to the ceiling, and she looked to where the finger was directed. She then looked back at him, understanding what he was implying.

"Oh, come on, Mick. Don't make jokes like that."

He knew it was either downhill or uphill from here; one of the two. "It's no joke, Jan. I've seen proof."

"What are you talking about, Mick? Have you lost your mind?" Janice couldn't hide her obvious concern for her little brother.

"I know this is hard to believe, and I didn't either at first. But I saw something I wasn't supposed to see early last week, and they had to tell me about who they are . . . I mean, what they are. They told me everything."

She shook her head slowly, still convinced he had lost his marbles. "What? What did you see?"

"Okay, I'll get to that, but first, just listen to me, and let me tell the whole thing before you make your decision."

"My decision? About what?"

He smiled a hopeful smile, knowing this was his chance to ease his burden of knowing something no one else did, something hard to carry by himself. He started with the confrontation Heidi had with Butch, then recalled for her each visit over the previous week at the house on Louisiana Street, describing what he had seen the afternoon they were forced to reveal themselves, then all the rest.

"Remember I asked you about those wigs last week? Well, when Rhett's fell off, I could see inside his head, and I saw his alien body, or part of it. And his arm was stretched out over fifteen feet, which wasn't human."

She was still staring at him like he was an alien himself.

"That started it all. They begged me to listen to their story. And I'm begging you now, because I need someone to help me understand that this is really happening."

"Help you? How?"

"Go with me the next time I go there." Seeing her shaking her head while inching away from him, he quickly added, "Listen to me, Jan. So far

I am the only person in the world who knows there are people from another planet living *right here*, in Cypress Springs, Texas, and it's driving me crazy not being able to tell anybody. And it's gotta be somebody I trust if I do tell.

"And you'll really like them, Jan. They are very intelligent, so far advanced beyond our knowledge, with no meanness about them at all. Yesterday, before I left, they showed me what they looked like under their human skin. It was beautiful, but hard to describe. You need to see it for yourself."

"What if they don't want anybody else to know about them. Didn't you say they're trying to stay hidden?"

"Yes, but it was their idea for me to bring someone, and Heidi knows about you, so they suggested you."

"She knows about me? How?"

"I don't know, but she knows everything about me too. I think she knows everything about . . . well, everything. Even about Kristi and me."

Janice, for the first time, seemed intrigued about the possibilities. Plus, she could see that her younger brother was really stressed over this. She held her fingers to her mouth, considering. Mickey then moved in for the kill.

"Think how exciting it will be when you hear about their world, and how they got here. You and I can talk about what they teach us. It'll be our secret."

She sighed, "Okay. But the first time I get the creeps, I'm outta there. And I might still change my mind."

"No problem."

"So, when do we go?"

CHAPTER TWENTY-TWO

We got problems. Seems like you were right."

"Right about what?" Winston Dunaway said, suppressing a yawn induced by a lack of sleep Sunday night. His wife, either growing increasingly suspicious of his philandering ways, or passing through another hormonal change in her female life span, had talked herself to sleep going over his shortcomings in the husband category. Only after she had succumbed at around three a. m., had he been able to drift off. And now this; a Monday morning call from Susan Taylor, aka Jane Robinson.

"Somebody did see us that afternoon in the kitchen. That's what." Her voice seemed to have a touch of satisfaction.

"Oh, shit!" He rose from his desk chair, circling the desk to make sure the door was closed. "How do you know?" he asked, his yawn slammed shut. He was instantly shocked awake.

"My son told me. He said kids at school are talkin' about it. Thought he was gonna be mad, since he's got a temper just like his dad, but he seemed more embarrassed than anything." She paused for a moment, as the man on the line stewed. "You know, Butch has been actin' strange lately," she went on. "Ever since early last week, he's been a lot quieter than normal."

"Maybe he knew about us back then?"

"Nope. He said he just heard about this on the way home from school, Friday, on the bus."

"Exactly what did he say?"

"That those kids were sayin' I was foolin' around with another man. In other words, someone other than Butch's dad. I felt bad for Butch. It really seemed to hurt him."

Dunaway could feel his stomach being tied in knots. There were two things he had to know to see how bad things were. "Has he told Jim about it?"

"No. And I told him not to. Said I'd talk to Jim, and that we would work it out."

"I hope to hell you won't tell him." Dunaway responded quickly. "That would be dangerous."

"You think I don't know that? Of course I won't tell him, but I gotta make Butch think I did."

"Did he hear anything about me being the other man?" he asked, holding his breath.

"Not that he mentioned, and I'm sure he would have if he'd heard that. Which means that so far, my name is the only one that's being sullied. You're in the clear."

In the clear, my ass, Dunaway thought to himself. If they had been seen, then whoever saw them had seen *him.* That person, who he was sure was a kid, wouldn't keep quiet about it forever. He or she, and he was also certain it was a he, needed to be found and dealt with.

"Listen now. I want you to talk to your boy, sometime when Jim's not around. Tell him he needs to find out who's spreading those rumors. And tell him that's just what they are—rumors."

"I think I convinced him of that."

"Okay, but still. We have to know who it was that saw us, in case things don't die down. If we find out who it is, then I can handle it."

"Handle it? You're not gonna hurt a kid, are you?"

"Of course not. They are other ways besides that. Don't worry about it. Just tell Butch to find out."

"He may not want to find out. Like I said, he's pretty torn up over it."

"Look. Tell him he'll be a hero, sorta like a detective. He'll be protecting his mom's reputation. That should do it."

"My reputation's already shot, like I told you. You're only concerned about your butt."

He wanted to scream back at her, but knew tact was called for. "No I'm not. I'm concerned about both of us. The sooner we find out who it is, the sooner it'll be nipped in the bud."

She made a couple of more objections, but he finally held sway. He simply had a lot more to lose. She hung up, with a meek, "I'll see what I can do."

When the line went dead, Dunaway sat still in his chair, wiping a hand over his bald head, now slick with perspiration.

CHAPTER TWENTY-THREE

Mickey had never had a real date. His innocent meetings with Kristi at the Dairy Queen, like the one he had departed thirty minutes earlier with an overload of homework excuse, could never be considered a romantic hangout; they were just that—kids, hanging out. A blind date was a term he was aware of, but was also only a possible event years into the future. But from his perspective, Janice gave the impression she was entering into such an arrangement with the family she was about to meet, which was scary; like walking into a lion's den.

During the relatively short walk from their house to 710 Louisiana Street the following Monday afternoon, she had threatened to turn back twice. She was waffling between being afraid of the unknown that awaited her, and a lingering, more fearful thought the creatures Mickey had described inside that house might actually do harm to her and her naive brother.

When Mickey bravely started to knock on the door, it opened as if by magic before his knuckles struck wood. Jan stepped back a pace, again growing weak-kneed.

"Hi, Mickey," Heidi said, her usual gentle smile in place. "And you too, Janice. It is nice to finally meet you. Please come in."

"Thanks," Jan said guardedly. The siblings entered, one with more than a small amount of trepidation. Mickey stood between them as Heidi continued her welcoming smile, while Jan observed her as she would a rare animal in a zoo. Heidi seemed to take no offense at the girl's suspicious appraisal as she walked around the two and closed the door behind them. Mickey, however, turned and frowned at the awkward rudeness, giving a quick shake of his head at Jan meant to quell the examination. She took notice and turned her attention to the room.

"Why don't you and Janice take a seat on the sofa, Mickey," Heidi said, as she stepped back and motioned. "The others will be joining us in a minute."

Heidi then positioned two of the dining chairs at opposite ends of the couch, evidently for she and Blanche. Mickey assumed Rhett and Jed would take their usual places in the wingbacks. Jan sat very close to her brother, their hips actually touching until Mickey slid a few inches away. "*Relax*," he whispered. "*It's gonna be okay*."

Heidi, of course, heard the exchange. "Yes, Janice. We want to assure you, just as we did Mickey that first time, you have nothing to fear from us. We will answer any concerns you have about us, and why we are here, and continue teaching Mickey, and you, if you want, about our world."

Jan's body, having remained taut with tension since entering, exhibited a noticeable release. She gave a forced grin and said, "Okay. I'll try." She pulled her knees up close by wrapping her arms around them. It was still a defensive posture, in spite of her loosening up. She didn't attempt to hide the fact she was eyeing the other supposed girl's appearance, concentrating on the thin, silky hair. She was evidently recalling Mickey's comment about hair pieces.

"The others wanted me to greet you alone if you agreed to accompany Mickey, Janice," Heidi was saying. "We did not want to overwhelm you at first. We also understand how unsettling this all must be for you, as it was for your brother at first. You, at least, had him to forewarn you. He had to learn our story alone. Our mission will be to make it worthwhile for both of you."

"I'm certainly willing to listen," Janice responded, her voice growing stronger. "And please call me Jan, if you don't mind."

"Of course, Jan." Heidi looked at the two sitting across, switching her gaze from one to the other. "To get started, before I ask the others to join us, are there any questions either of you have now?"

Mickey lead off. "I've brought Jan up-to-date on what y'all have told me so far, but something you just said reminds me, we never have really talked about why you have come to Earth in the first place."

"Certainly. An excellent question, Mickey, and one we expected earlier. The two short answers are exploration and observation. Are you familiar with the television series titled, *Star Trek*? It ended shortly after you were born, and while Jan was also still very young."

"Yeah," Mickey replied, with his sister nodding agreement. "It's on all the time now in re-runs. I like it okay, but doesn't compare to *Star Wars*."

"We agree. The technology available to current movie makers is far superior to that which existed in the decade of the sixties, when the series was being produced. If you recall the dialogue at the beginning of each show, the mission of the travelers on the Enterprise was to explore new worlds within The Milky Way, and to search for other forms of life and civilizations. Of course, as we discussed during your last visit, the beings they discovered were entirely fictional, since we have witnessed no such life forms while on our journey through your galaxy.

"We were gratified to detect life on your planet when our scientists did the same as Captain Kirk and his fellow travelers attempted to do. It took many Earth years for us to ultimately arrive, but our purpose in being here is to become accustomed to the conditions on your planet in the hopes we can share our knowledge with each other in the future."

As Heidi spoke, Mickey was struck by her words and demeanor. She was speaking as an adult now, no longer a fellow fifth grader. A superior intellect was shining through. He would likely never again view her the same, and he wasn't sure he liked that idea. And he was starting to believe the girl may not be a female at all, which was beyond weird.

As Heidi started to continue, the other three inhabitants entered, taking their seats in the manner Mickey had anticipated. As Heidi introduced them, they each smiled at Janice in greeting. Mickey said, "Hi", while Jan simply stared, focusing on the tops of their heads.

"Unfortunately," Heidi said, "as Blanche explained earlier to Mickey, it may be some time before we can actually share our abilities with Earth's natives, due to the fear of the unknown she referred to. But we have to start somewhere, even if it takes many years to remove those doubts."

Blanche leaned in to speak. "Which means you two, Mickey and Jan, should consider yourselves the first step in our acceptance by Earthlings. For if we can, over time, convince you we have no other motives than those Heidi has explained, then perhaps others will give us the opportunity. However, we realize it will be a long process, and we are under no time constraints."

Mickey snuck a quick glance at his sister. She appeared to be absorbing it all pretty well. Removing her arms from around her knees, she actually gave a narrow grin in response to Blanche's words.

"Can I ask a question now?" she blurted out.

"Of course, Jan. Anything you wish."

"Well, Mick told me how you got here, and the night we saw those lights in the sky, which he said were your arrival. I think I understand that, as amazing as it sounds. But, what I can't figure out is how you manage things here, in Cypress Springs, or wherever you go." Jan paused, putting her thoughts in order. "I mean, how do you manage to find a house? Pay your bills? Buy food? Do any of you have jobs?" There was a touch of suspicion in her tone.

The group of inhabitants shared knowing smiles. Then, Rhett responded. "Amazing questions, Jan. Mickey explained how your mother uncovered our method of choosing our Earth names. It appears you have the same curiosity and intelligence she has. The answer to all of those concerns is the same. We have funds available to us to survive a simple existence. However, food is not an expense for us. We do not possess digestive systems similar to yours, requiring a special diet, as you might refer to it, to sustain ourselves. We do not require much nourishment, but what we do require, we produce ourselves."

The siblings could only sit and stare, somewhat dumbstruck by what Rhett had explained. Mickey's next thought was a mental picture of Heidi in the school lunchroom, watching her eat; or maybe, not eat. She must've been dumping her tray somewhere, or possibly sneaking her own meal in somehow. This was becoming so weird, he wondered how Jan was coping with it.

As if in answer to his concern, his sister pressed on. “What about the rest? The other expenses? How do you get the money?”

“The answer to that is that it has been provided by someone who visited Earth before we arrived. That individual is no longer on your planet, having departed when we left orbit and descended to Earth. The citizen of our planet whose place we took had worked among humans for a few years, earning a living, while amassing an adequate amount of funds for us to manage well. We met with him on our mother ship before we descended, and that individual returned to our planet with that group.”

Now Mickey and Janice were stunned into complete silence. Mickey felt the sudden disappointment of discovering he was likely not the first to experience the thrill of meeting extraterrestrial beings face-to-face. As he tried to adjust to that knowledge, Jan continued. “Where did that person . . . or being, live and work?” She shook her head. “I’m sorry. I apologize for not knowing how to refer to those from your planet.”

“That is quite alright, Jan. Whatever you prefer is fine with us. As we explained to your brother, the name of our world does not translate to your language. Otherwise, we would attempt to pronounce it. We understand it is difficult for you. However, in answer to your question, we had rather not divulge the location of our predecessor’s assignment. We may be utilizing it again in the future. By the way, he was here alone, which made it simpler for him to be absorbed into society.”

Rhett seemed to recognize the disappointment written on Mickey's face. "No need to worry, Mickey. That first individual never divulged his identity to anyone, which means you *are* the first to know. Even in Michigan, where we explained we had visited before your community, we had no interactions with the citizens."

That fact did help some, but something similar to a hollow feeling still remained. Part of what Rhett had just said reminded him of something Heidi had said the last time.

"Heidi explained to me that if I hadn't agreed to hear your story, you may have had to relocate. Are there other reasons that might happen?"

Jed finally joined in. "Hopefully not, Mickey, however, it is impossible to predict things such as that. We can only trust that you have convinced Jan how critical it is to us for our mission be kept between us at this time."

Before her brother could answer, Jan jumped in. "Yes he did, and I completely understand. This has all been amazing, but . . ." Her voice tailed off, as if afraid to continue.

"What is it, Jan," Jed said with his hands spread apart. "Feel free to ask or say anything."

She looked at Mickey before going on. He gave her a doubtful look, unaware of what was on her mind.

"Well, it's not that I'm questioning anything at all about what you've told Mickey, and now me. But I was wondering if I could possibly see something he saw at the beginning. I'm feeling kinda stupid even asking, but it might help."

Jed now showed a wide smile, as he leaned back in his chair. "Ah, yes. I think this may be what you have in mind." His left arm began to silently extend beyond where the flesh-like skin ended. It grew and grew, with a scaly hand at the end coming toward the wide-eyed girl, as if launched. It switched course when it neared her, coming to rest on Heidie's narrow shoulder. Jed allowed the hand to stay there for a few seconds.

Jan sat back in wonder, not believing what she was seeing. She looked over at Mickey, who was laughing, first at her, and then with her.

"How is that, my dear?" Jed asked, now starting the retraction.

"That is so cool!" she exclaimed, barely able to contain herself.

Blanche broke in to her enthusiasm. "I believe we have reached another breaking point. We must give you two time to assimilate what has been learned so far." She turned to Janice. "As we have told Mickey, please bring any other questions you have to our next meeting. That is, if you want to continue with us."

"Are you kidding? You couldn't keep me away."

Walking home minutes later through signs of early dusk, shuffling through the red, yellow, and brownish leaves beginning to gather on the sidewalk, the two giddy siblings would stop every few steps, discussing what had just transpired. Jan did most of the talking, in an animated fashion. The dam had been breached, and the knowledge of what she had thought impossible was washing over her. Twice, Mickey had to warn her to keep her voice down.

"Let's go back tomorrow," she insisted.

Mickey knew he had to reel in her thirst for more. She was in the very same position he had been in a week earlier, and he knew how easy it was to become overwhelmed by too much, too soon. By the time they reached the front steps of their home, he had talked her into Wednesday rather than Tuesday. He also reminded her that their faces must not reveal anything to Jessica. The poor woman had enough problems already.

CHAPTER TWENTY-FOUR

With the successful introduction of Jan to the 'Jones', and vice versa, Mickey enjoyed a deep, deep sleep Monday night. The same could not be said for his sister. She had to endure the brain games he had one week earlier, her head filled with what she had been told, and the unworldly demonstration by Jed. At breakfast, she was a zombie, staring down at her cereal bowl. In her defense, she had held together during the night before, deflecting questions from Jessica about her daughter's mood, which was more hyper than normal.

Mickey kept a close eye on the girl, who gave an occasional glance in his direction across the table, accompanied by a sleepy grin. He was aware the rest of the day would be the real test for her, staying mum about the amazing things she now knew; things a normal thirteen year old girl shouldn't be asked to conceal. But she had promised him she could pull it off. Based on her appearance and energy level, it was definitely a good thing they were taking a break tonight. Later, as she headed out the door to join her car pool, he stood on the landing, ready to head to his bus stop. He got her attention as she climbed into her friend's car, mouthing a silent, *"promise"*? She nodded in affirmation. It seemed strange to him, but the normal routine between the two of them was now turned on end; he was now the monitor of *her* behavior.

The first half of the school day passed in typical fashion, with the only ripple being the continuing icy relationship between Mickey and Trey. A stare-down occurred during the one class they shared—social studies. Their cold war apparently was going to last for a while. Halloween was on the horizon, only a couple of weeks in the future. The four of them: Mickey, Benny, Trey, and Roby, assuming Senor Pepe's could spare him, had made the rounds the last two years together, heavier on treats than tricks. It was doubtful Fulghum would be joining in this time, but with everything else going on in his life, Mickey couldn't waste time worrying about hard feelings. Of course, there was no escaping the fact those feelings were due to Trey's treachery, and of course, *his own* big mouth.

He, Benny and Roby had just finished lunch, and were ambling outside to the familiar patio, when they caught sight of Trey leaning against a brick wall in the corner where the outside area and the main building met. He was being hemmed-in by the hovering form of Butch Robinson, who appeared to be in the process of jamming a stubby finger into the other boy's chest. He was trying to keep his voice down, but a few threatening words escaped, filtering through the voices of others outside.

Mickey immediately knew the purpose of the threats could only be one thing; the source of the embarrassing news regarding Butch's mother. "Oh, crap!" he said, a little louder than he intended. "I bet he's trying to get Fulghum to give up his source about his mom."

"Exactly what we thought he might do," Benny added. "We warned him it might happen. Think we oughtta go help Trey before Butch does some damage?"

As they watched, Trey tried to make his way around the bigger boy, but was stopped with a meaty hand, and thrown back into the wall.

"If we're gonna," Roby said. "we'd better be quick about it. Robinson looks like he means business."

"Yeah," Mickey said. "Let's go." He was sensing a heavy dose of guilt as he started for the scene of what had signs of a bad situation. This was all his fault. His decision to show up his friends by proving he knew more than they did about something that wasn't at all important had put this situation in motion. Benny and Roby trailed behind him. He wondered if they shared the thought that Butch might very well handle all three of them, plus Trey. He was certainly capable. The thought also struck home that by entering the stand-off, Trey might feel inclined to identify him as the source to protect himself. But there was no choice; he couldn't stand by and see harm done.

When they were halfway to where the two boys were squared off, Mickey saw a female form exiting the lunchroom. He was horrified to see the girl everyone else knew as Heidi Jones heading straight for Butch and Trey. She was moving silently, using her usual stealthy motion. Trey saw her first, looking past the bigger boy who now had him by the shoulders, pinching them together. Butch turned to follow Trey's eyes, then his own orbs grew wide at the recognition of who approached.

She stopped a few feet away, then leveled a friendly smile in the direction of the two boys facing off. "Hello, Butch," she said in an even tone. "You too, Trey." The other three friends stopped in their tracks, watching and waiting to see what was going to happen. Mickey was terrified. He knew that if she used the same trick as the previous week, it would be seen by the other kids on the patio who were now turning their attention to the drama. It would be hard to explain that away. He wanted to yell her off, but was fairly certain she would simply ignore him.

He watched as Butch released his hold on Trey, who appeared to be ready to collapse. The bigger boy now showed fear registering in his own wide eyes. He then attempted to put a scowl on his round face to show courage, aware others were watching. Heidi then moved one step toward him, raising an arm in his direction, and that was all it took. Butch took a last look at Fulghum, said something to him, then disappeared around the side of the building. Heidi quickly did an about face, re-entering the lunchroom. Only then could Mickey release the breath he had been holding in.

The three friends approached the still shaking boy huddled in the corner. Mickey wanted to be sure about the purpose of Butch's attack. "What was that about, Trey?" he asked.

"Nothing. Just leave me alone." The be-freckled red headed boy started to ease himself away from the others. But Mickey had to know.

"Did you tell him, Trey?" Mickey asked, moving to block his path to freedom.

"Tell him what?" The tone was meant to be defiant, but lacked steam.

Remembering the boy had been a close friend less than two weeks earlier, Mickey softened his tone. "You know what. Look, I've already told you I take most of the blame for this whole deal. I just need to know where things stand, is all."

Trey stood up a little taller, glaring back. "That's right. And yeah, he did want to know who told on his mom. He found out I had mentioned it on the bus, and wanted me to rat out who saw it. But I didn't tell him. Okay! Now leave me alone."

He pushed Mickey to the side, and walked back inside the building.

The other three watched as he hurried off. "Wonder what Butch said to him before he took off?" Benny asked.

"No idea," Mickey said, but he had a good idea—a frightening one.

CHAPTER TWENTY-FIVE

Wednesday morning dawned with an overcast ceiling hanging over Cypress Springs, which should have served as an omen that something bad was brewing in Mickey Foreman's life. But there were no outward signs of the potential for worry at first. One not so obvious sign came during social studies class, with the absence of Trey Fulghum. Curious about the no-show, Mickey asked around. No one knew for sure, but a few said it could be a flu bug, an early version of what the so-called medical experts on tv said was to be an active cold and flu season. Some of his classes the rest of the day were regular cough-fests, so there was some credence to that possibility. Absent that conclusion, he thought the answer might lie in the case of Fulghum not wanting to face his classmates after his skirmish with Butch Robinson on Tuesday, and his rescue by the pencil-thin girl whose mere presence had put fear in the bigger boy. It had been an embarrassing few moments for the normally brash boy.

Butch Robinson *was* in attendance, seemingly taking every chance he had to cross paths with Mickey, each time with a glowering expression that appeared to be half malice, half satisfaction. Given what had transpired the day before, there seemed to be no reason for that attitude, so it was a puzzle weighing on Mickey. He did his best to shrug off Trey's absence and Butch's smarmy demeanor, feeling there was no connection between the two. Besides, he was looking forward to the afternoon, when the second round of meetings with Jan and Heidi's "family" were on his schedule. When his sister had hurried out into the gloom of the morning that day to join her car pool, she had given him a thumbs-up, which he took to mean she was ready to learn more. He felt good about that, mainly because he saw her back on his side again; no more snarky comments.

The voice Winston Dunaway hated hearing spoke into his receiver. "It was Mickey Foreman."

"Who?"

"He's a friend of Butch's, or at least he used to be," Jane Robinson said sleepily into the phone. "He's the one that saw us that afternoon."

"Why is he a used-to-be friend?" Dunaway really didn't give a crap why, it was just a question to give him time to think, now that he had a name. It was early Wednesday morning, early enough to fear Jim Robinson might still be home. But since she was talking freely, he assumed the guy was at work.

"I don't know that. All I know is Mickey used to come to our house some. But not lately. He's been in here enough times though to feel like he knows the place, so it makes sense he mighta snuck a peek in. You know how curious kids are."

No, he didn't know, not having any of his own. "What is he, some kinda peeping Tom?"

"Course not. He's a good kid, near as I can recall. He was probably just crossin' through my back yard. Lives just a few houses away."

"How did your boy find out?"

"Well, he says he had to get a little rough with the boy who knew about Foreman. Got interrupted by somebody during lunch period, so waited for him after school and showed him what's what."

"And you were worried about *me* hurtin' kids?"

She gave an exasperated sigh. "You said for him to find out, didn't you?"

He let her smart reply pass by. Hell, he couldn't wait for this whole thing to end, if it ever did. "Yeah, well, but we had to know. What else do you know about this . . . Mickey kid? Where does he live, exactly?"

"Down my street, then a right on to the next corner. Guess that would be the corner of Sabine and . . . now let me think . . . oh yes, that would be Tyler Street."

"Which corner? Southeast, southwest, or what?" Never ask a woman for directions, Dunaway recalled too late one of the truisms he lived by.

"Oh, hell, I don't know. You can look her address up in the phone book."

"Look who up?"

Another sigh, this one more impatient. "The boy's mother. She's a widow. Name's Jessica Foreman."

He took that as a good sign. No daddy to worry about if things got messy. "Okay. I'll find it. Tell your boy he did good. Just make sure he doesn't mention all this to Jim, or anyone else, for that matter."

"So, what are you gonna do, now that you know?"

"I haven't thought through that yet. But don't you worry, I'll come up with something." His mind was already busy shuffling ideas in his head.

One thing was true, he needed to come up with something, and quick. He knew how these things worked. Either her slug of a son would talk, or the kid he got the information from would, spreading news about his involvement across the small city. But wait. Nothing had been said about *him* being identified—only the woman. Maybe he was still in the clear. Hell. He couldn't think that way. This Foreman kid knew, so he still had to do something. As he mulled ideas, she interrupted his thoughts.

"Listen, Jim told me last night he's gotta go to Nacogdoches for a couple of days, either later this week or next. If you really want to slow down, give me one more date and we can talk about it."

A part of his rear anatomy suddenly puckered up. *Just when he figured he might be getting out*. He knew how that talk would go.

"Well, give me a call when you know his schedule, and we'll see," he said, hanging up before she could reply.

CHAPTER TWENTY-SIX

The late afternoon sun was starting to sneak between semi-dark clouds when Mickey and Jan prepared to enter 710 Louisiana Street. A few moderate showers had left puddles here and there, causing a few sidesteps along the way, but nothing could stench their buoyant mood. They had been freed up to arrive a little earlier than usual, due to Kristi showing signs of the bug. Catching him before leaving school, she explained while sneezing into a Kleenex she was going straight home to bed. No D Q today.

Jan mentioned during the walk over she had a question or two, which was fine with her brother. He looked forward to having a chance to sit back and listen and observe for a change. His sister had a quality only females seemed to have—the ability to dig into a subject until all parts are exhausted. Their mom had that talent in spades, so maybe it was an inherited trait.

Heidi opened the door as usual when they reached the top step, ushering them in. Everyone, with the exception of Jed, was back in place, which left one chair empty.

"Jed will be with us in a few minutes," Blanche said, indicating the vacant seat. "He is working on some of our communication equipment in a back room."

Her explanation made Mickey want to see whatever was in that back room, but he thought it might be pushing things to make that request. If there was equipment like that back there, how far did it reach? All the way to their home planet? Could that even be possible?

"Well," Blanche began as the two guests settled in to the sofa. "Where would you like to begin today?"

Jan opened up with her excitement showing. "The thing Rhett did with his arm before we left Monday, is that something like the weapon Heidi described to Mickey last week? The electric magnetism, or whatever it's called. Are you all born with that ability?"

"It is actually referred to as electro-magnetism, which is an energy source very common here on Earth. Its existence has been known for over a century in Earth time measurement. However, the mobility of our arms, in fact all of our extremities, has evolved within our bodies over time. And, we certainly do not wish to think of our inner defense mechanism as a weapon. As Heidi explained to Mickey, we can indeed bring it to the surface when needed, as you can your abilities by blinking your eyes, or shrugging. Those are things you can do, just by willing them. We were not happy to learn she had used her power when Mickey's friend seemed to be a threat to her, but she explained her reasoning for doing so."

"From what I heard, he had it coming," Jan said.

Mickey wondered if Heidi had told them she had come close yesterday to zapping Robinson again. Whether she had or not, it wasn't his place to mention it now.

"So," Jan continued, wanting more. "Those powers, or whatever you want to call them, have developed over the years? Passed down from generation to generation?"

"Exactly. Think of animals in your world that have characteristics that have evolved over the years for protection, both from predators and the elements. Snakes have developed scaly skin for protection. Some creatures can possess the ability to change colors to blend in to their surroundings. And of course, those that reside in colder climates manage, over centuries, to grow thick coats for protection."

"Oh, yeah," said Mickey. "We've studied that in biology."

"Of course," Heidi said. "And, even humans can evolve. You are all much taller than your ancestors were, as you can see from older photographs. Intellect can be sharpened when two people with well developed brain functions mate, creating children with abilities to excel in school at a high level. There are many such examples."

The room grew silent for a moment as Mickey and Jan absorbed it all.

Blanche then filled the void. "We are all beings, Jan. Whether we are Earthlings or from our galaxy, animal or perhaps plant life. All living things evolve, however from different beginning points. Something to keep in mind is that our planet's civilization is many centuries older than yours, in fact, perhaps thousands, which has allowed us to develop to a point beyond Earth's."

Jan nodded slowly, as if that all made sense. Then she had another concern.

"This is changing the subject some, but Mickey explained to me about the pod, or whatever you call it, that landed you all here. Is it really hidden outside town somewhere?"

"Yes it is," said Jed, answering the question as he entered the room from the mysterious back regions. "Hopefully, in a safe place. We refer to it as a module, by the way." He was carrying something in one of his hands, a small gray metallic-looking object.

"Well, what I was really wondering is, how do you get around town to go shopping, or whatever. I can't imagine you go get that . . . module, every time you have to go anywhere."

"Indeed we do not, and fortunately we do not require many excursions. As Heidi explained, we do not require visits to food stores, and we do not object to walking to most places."

"Do you even have a car? I didn't see one in your car port."

It was Rhett who answered. "We do not possess one, but have found it a simple matter to rent one if it becomes necessary. I have been able to obtain a Texas driver's license, which is a requirement to rent. Again, that is a rare occurrence. I did the same in Michigan. We have utilized that service enough now to simply call the dealer, and they will deliver a vehicle to us."

Jan looked around the room. "How do you call? I haven't seen a phone anywhere yet."

Rhett turned toward Jed, who held up the device he had in his hand. "This is one of our communication devices. It serves as a telephone. Mickey used it to call you the first time we began these sessions. I will explain to you how it functions at some point."

"Oh, yeah. I remember now. It was amazing how clear it was."

Satisfied with the answers to that puzzle, Jan hesitated before posing her next question, wondering how far they would agree to go to quench her curiosity. Well, they had said, "anything" was fair game.

"There was something else Mickey told me about, it was something one of you did the last night before I showed up. He said you showed him what you looked like in, well, what I guess you would call your natural state." She could sense them all staring intently at her, anticipating her actual request. It made her feel uneasy, wishing she had never started in this direction. But since she had, "Would you do the same for me?"

The four inhabitants of the house all rose, gathering near the dining room table, conversing with each other. Evidently reaching an agreement, Blanche emerged as the spokesperson. She addressed the two visitors. "Heidi will again remain as she is. We continue to think Mickey would rather she maintain her human appearance. We other three will gladly give you a small sample of one part of our life on our planet."

With that, they disappeared into the kitchen area, much as they had the night one had revealed themselves to Mickey. Heidi rose from her chair, closing the blinds and curtains covering the windows in the living room and dining area, creating a somewhat darker room.

Music began to emanate from somewhere behind the doors leading to the kitchen. It sounded strange at first, then Mickey and Jan realized it was familiar, but not the sort of music they would normally pay attention to. It was the weird music they were forced to listen to late on Saturday nights when they stayed over at their grandparents' house in Texarkana. The Knoxs loved it, but even their mother thought it was too dated. It was the so-called Champagne Music of band leader, Lawrence Welk. It began to waft through the house as a large green-hued mass suddenly appeared through the door away from the dining area. It hovered a few feet off the floor for a second or two, pulsating to the bouncy tones of the old-fashioned music as it threw off sparks. Then, it separated into three distinct shapes, yet shapes that were indistinct in their own designs.

As the siblings on the sofa stared in wonder; the boy smiling, the girl with eyes bulging and mouth hanging open, movements that could only be classified as dancing circled through the room. Round and round they twirled, keeping perfect rhythm to the sounds of Welk's orchestra. Gradually, Jan could make out the human-like forms within the fuzzy green shapes that Mickey had seen, rapidly changing their outlines as they floated across the room. At one point, they rose to the ceiling, flattening out to circle the room from above. The two sets of eyes on each being opened and closed as they navigated over to hover right above the amazed humans. The two on the couch had to lean back to view them full on. Jan would later tell her brother she saw the same crescent smiling eyes he had the time before.

Mickey thought the relative darkness of the room added luster to the bright sparks, enhancing things. But he also knew that those bright flares were the real reason Heidi had closed the window coverings. Passers-by would certainly notice, and wonder. He took a quick look at his sister, seeing laughter mixed with tears of total wonder; she had her legs up under her body, sitting as far back on the sofa as possible.

As the tune began to come to an end, the three shapes reunited in one large blob again, and as quickly as it had appeared, instantly floated from the room.

While they waited for the performers to regain human forms, Heidi felt an explanation was due. "That was not a typical occurrence in our daily life. I imagine they were again showing out, as you would call it, to a certain extent." If it were possible for her to exhibit embarrassment, her narrow cheeks would be blush red, but they remained ashen.

It took a minute for the effects of what she had seen for Jan to calm down. Leaning against the back of the sofa, she appeared to be washed out.

"I don't know what to say," she murmured. "That was absolutely the most beautiful thing I have ever seen. It blew me away."

After the others re-entered the room and things had settled down, Mickey and Jan thanked the participants for the fantastic demonstration. Any further questions the two had for their hosts would certainly wait, so the meeting soon came to a halt, with the promise on both sides to reconvene soon.

The walk home was more subdued this time. Jan now appeared fully on board with the possibilities she and Mickey had ahead of them.

Wrapping an arm around her brother's shoulders, she said, "Thank you, Mick, for talking me into this. To think I thought you'd lost it, and how close I came to blowin' your story off. It's a shame we have to keep this a secret."

Stepping back from her a step, he reminded her, "But we have to, Jan."

"I know. Just wishing."

They had been in the house on Louisiana Street longer than they realized, and were surprised to see the Toyota in its usual spot at their house. Jessica was home earlier than she had been lately. They took no notice of the other vehicle, the one parked around the corner of their house on Tyler Street.

Erasing the emotional residue of the late afternoon's happenings from their faces, they entered the house, where Mickey Foreman came face to face with his own personal nightmare.

CHAPTER TWENTY-SEVEN

They entered through the front door as nonchalantly as possible, with faces like blank slates. Mickey was concerned their mother might be suspicious of where the two of them had been together; they were not prone to hanging out with each other, especially lately.

A few steps into the house, they came to an abrupt halt. There were voices coming from the living room/den, one of which belonged to Jessica, but the other, a male voice, was unfamiliar to both. They looked at each other, shrugging. Unsure of what it meant, they slowly approached the arched opening into the room from the hallway. Peering in, they could make out Jessica, sitting on the worn out sofa. She was looking in the direction of the recliner, set in the upright position, with its back facing them. A pair of legs in khaki pants, legs crossed, ankle-high black leather boots on large feet, protruded from the chair.

"Oh, there they are now," Jessica said, her face lighting up. "We were just talking about you two."

The man sitting in the chair put a hand on the arm of the chair, while leaning around to look. He then stood, as if ready to give a polite greeting. "How're you kids doing?" Chief Winston Dunaway asked, looking at the two of them, but with his eyes trained mainly on the boy.

Mickey felt as if the room was instantly compressed down to the point that there was only the man and him standing in it. The floor seemed to be sinking, walls falling in on him. Noticing the man's holstered weapon tied to his belt, he looked for handcuffs attached there also, but thankfully, saw none. He was totally unaware of the presence of either his sister or his mother, but then he could hear Jessica speaking in a voice that sounded far away, as if filtered through a long tube of some sort.

"Kids, Chief Dunaway came by to make sure things are okay. Wasn't that nice of him?" was what Mickey thought he heard. But he couldn't be sure, since he couldn't remove his eyes or his thoughts away from the man, who continued to stare at him. The law man's height seemed to reach the ceiling, his bald head reflecting the overhead light. His eyes were weird. The pupils appeared to be floating high on a sea of whatever that while stuff was called, giving him a sinister look to add to the threat of him actually being in this room—in this house. Those pupils were the blackest black Mickey had ever seen, which only added to the fear factor.

As the reality of the situation settled in and the room expanded to its normal size, he was aware Jan was staring curiously at him. She then looked toward her mother. "Why shouldn't things be okay? What's happened?"

"Well," Chief Dunaway responded instead. "Turns out it was a false alarm. Somebody called in saying they saw somebody actin' suspicious outside your house here. We've had a few burglaries in the area, and so, well, I was in the neighborhood when the call came in. Heard it on the radio. Since I was close, decided to check it out myself."

"Mom?" Jan followed up. "Did you see anything?"

"No, honey. I'm sure what Chief Dunaway said is right. He did a complete check outside the house, bushes and all. Nobody there," she added perhaps too cheerfully.

"It never hurts to be extra careful," Dunaway said, turning his gaze directly back on Mickey. "You never know who might be right outside your window, taking a peek inside." He then addressed Jessica, furrowing his brow. "You know, I never forget a face. Didn't I spot you folks at Senor Pepe's a couple of weeks ago?"

She smiled a warm smile that her stressing-out son was horrified to see, but there was nothing he could do or say about it. He recalled vividly the leering look the creep gave in her direction as they left the restaurant.

"Yes, you may have," Jessica said. "We go there often."

"Good place to go. Lotta food for a decent price. The missus and I go there a good bit too. Maybe we'll see you there some time."

"Perhaps so."

Dunaway sucked in a shallow sigh, puffing his chest out, saying, "Well, guess I'd better be headin' home. Looks like everything's okay here. Nice to meet all of y'all. If you ever need anything, just call the department. That's what we're there for."

As Jessica walked him to the door, the man gave one last glance down at the boy who had spotted him with Jane Robinson that miserable day. And if he was reading the boy right, and he was pretty sure he was, he felt certain the boy now knew that *he* knew. Maybe that would be enough to keep him quiet. But it wouldn't hurt to keep tabs on him for a while. He walked down the walkway to the sidewalk, then turned the corner to his car, feeling pretty good about things.

Jan hustled Mickey back to her room as soon as the man had left the den. Closing the door, she wasted no time, afraid Jessica might look for them after the man had left. "It was him, wasn't it?"

"What're you talkin' about?" But he knew.

"He was the man with Mrs. Robinson. Don't deny it. I could see it on your face, how you reacted to him."

Mickey knew at once it really didn't matter anymore. The man somehow knew it had been him. Butch must've worked on Trey some more. *Crap.* He collapsed on her bed, putting his head in his hands. "Yes. It was him. And what's worse, I think he knows it was me that saw him."

"I'm Sure he does. I could see it in his face too." She took a seat besides him, putting a hand on his arm. "You know, I think the only reason he came here tonight was to scare you. There was no prowler. He made that up."

"Well, he succeeded then." Mickey was aware of Jan's hand feeling warm to the touch, sensing a smaller amount of the heat when she had hugged him while waking home. Her nose had a trickle beginning to leak out. She looked tired. "You okay?" he asked.

She gave a sideways look. "I guess. Why do you ask?" The words were barely out of her mouth when she fell into a short sneezing spree. She wiped her nose with a sleeve of her blouse. "Oh no," she said. "I hope this isn't what I think it is. Carole Yeager sits right next to me in trig. She was sinking fast today. Must've caught whatever she has."

While they considered her possible pending bug, Jessica knocked, then entered. She had an odd expression on her face, looking from one offspring to the other.

"Any reason why you two acted so rude to Chief Dunaway? Y'all barely said boo to the man. And Mick, you looked like you had seen a ghost."

Mickey had to think fast, or else tell her what he knew about the man. He didn't want to saddle her with that burden, so he came up with, "I'm sorry, Mom. I didn't expect to see him in our house. Just caught me by surprise. And I was afraid something bad had happened when I saw who it was."

She didn't seem to be completely buying it, wearing a look that said she thought the boy might be in trouble of sorts, although that would be unlike him. She was about to say something else, when Jan emitted what could've only been a choreographed cough, accompanied by a sneeze that likely wasn't. Jessica peered at her closely.

"Oh, honey. Are you coming down with something?" She touched the girl's forehead, finding the same low grade fever Mickey had felt. "Yeah, I think you are."

"I might be. Carole coughed right in my face today. She's got some kinda bug."

Jessica had heard enough. She told Mickey to get off the bed and leave the room, "Right now. Can't have you both sick."

As he did as ordered, he took a peek back at Jan. She gave a quick wink just before he closed the door.

Thanks, Sis, he thought to himself.

CHAPTER TWENTY-EIGHT

Thursday morning, and Jan's condition had grown worse during the night. Jessica did all she could to keep fluids in her, or else within reach on her night stand. Tylenols were taken to the extent the dosages on the side of the bottle recommended. The chills that always came with the yearly strains of flu viruses had the girl alternating between sweats and shivering fits, as she kept bundled up under three layers of bedding. Deciding to stay home to see to her daughter, Jessica, after driving Mickey to school (she convinced him the bus would be a prime breeding ground for whatever was making the rounds), she hurried home to call her boss. It was a bad time to miss work, with the new project underway, but he made a good attempt at being understanding.

Mickey experienced a largely sleepless night himself. The surprising appearance of the police chief in the house, following the spectacular viewing of what he was now labeling the "alien dance", if only to himself, since the only other human the phrase would mean anything to could care less at the moment, again had his brain conflicted. The two events, occurring within thirty minutes of each other were almost too bizarre to comprehend. During the drive to school, he was captive to his mom's desire to probe, asking if there was anything going on in his life that was causing him to act so preoccupied lately. She was concerned.

"No, Mom. Everything's fine. Just tryin' to get straight with Kristi again. Guess it's okay now." *Big fat liar*. He hated to lie to her, but it was better for her sake than telling the truth, at least that was his reasoning.

"What did you think about Chief Dunaway? Nice man, didn't you think?" she asked, looking over at Mickey.

It took a good bit of will for him to answer with a false opinion. "Sure. I guess so." *Crap. He wished he could've ridden the bus.*

She wasn't done yet. "You and Jan were out together yesterday afternoon. Guess the cold war is over?"

He knew he had to be careful with what he told her about that change. "Yeah. She went to Heidi's with me. Monday too. I wanted to show her Heidi was okay, and I think she really liked her."

"Oh? Well, that's nice." She took a sideways glance at him. "You've been spending a lot of time over there, haven't you?"

Squirming in his seat, he answered, "I guess. She's really helped me with my school work. Seems to know a lot about everything." He knew he had to add more. "That's okay, isn't it? I mean, they are all real nice. You don't need to worry about me being there."

"I suppose it's alright. Just don't be a burden to them. And don't forget your other friends—and Kristi, of course."

"No worries there, Mom."

Thankfully, the school building was soon in sight. The car had hardly come to a stop when he opened the door to extract himself. "Thanks, Mom," he yelled back.

Kristi's bout with the bug wasn't any better either; she was again missing from action at school. As much as he enjoyed meeting with the girl almost daily, he had so many other things on his mind it really didn't bother him to have a break, even though it was for a bad reason.

Trey Fulghum *was* in attendance, so it couldn't have been the flu; no way he could be over it so quickly. It took only a brief sighting of the redhead to figure out it was something else. A shiner could be detected under his left eye; noticeable scars ran down the side of his face. In the classroom they shared that morning, Mickey overheard Trey explaining to someone he had crashed his bike, but he didn't believe that for a second. As much as he regretted Fulghum's inability to keep their secret, the guilt Mickey felt over putting one of his closest friends in jeopardy by disclosing what he had seen was a heavier burden. He had to make it right—or as close to right as possible.

Catching up to Trey as they entered the cafeteria, he managed to pull him aside, although he resisted at first. "Leave me alone, Foreman," he said, pulling away.

"Please, Trey. Just give me a chance to apologize again." Up close, he could see scrapes on the left ear lobe too. "Butch did this, didn't he?"

"What do you think?"

"When?"

Trey seemed to realize there was no point in concealing his shame and embarrassment. "Day before yesterday, after school was out. He managed to hide on a back seat on my bus. Got off at my stop and followed me. Said he knew that girl couldn't help me this time, so he . . . well, you can guess the rest." He shuffled his feet; head down.

Mickey felt even worse now. "Well, it's okay if you told him it was me. Just sorry it worked out this way."

Trey trained his good right eye on Mickey. "He wanted to know who the man was, and I told him you never told us."

"Yeah, and I'm glad I didn't." As soon as the words left his mouth, Mickey felt confused. If Butch didn't know it was the chief, then how did . . . ?

"Trey, did Butch say he was going to tell his mom it was me?"

He shook his head. "No. Why does that matter?"

Breathing out a shallow sigh, he answered, "It doesn't. Just trying to figure something out."

"What?"

"Nothing important." But Mickey's mind was reeling. As they went their separate ways, he yelled back, hoping in one small way to put things back in place, how they had been before it all came apart a few weeks earlier. "Hey, we on for trick and treating next week?"

"Maybe. We'll see."

Later, as Mr. Jabroe, the quirky biology teacher, tried to gross out the entire class by describing how they would dissect a frog on Friday, Mickey drifted off into his own world of problems.

The only way Chief Dunaway could possibly know it was him, and he was dead certain Jan and he were right about that, was if Mrs. Robinson told him. Butch must've told her after he beat it out of Trey. But he was pretty certain the woman didn't share the man's identity with her son. There would be no point in that, plus it would put Butch in potential danger too. What that meant was that now four people were aware of the identity of Mrs. Robinson's boy friend—or whatever he could be called. Besides himself, there was Butch's mom, the chief, and now Jan. Of that group, the only person he feared at the moment was Chief Dunaway, and the man no doubt had him on his radar screen.

He suddenly realized how much he looked forward to his meetings with Heidi's amazing family. They were becoming the calm in the storms his life had sailed into. They never complained, like most people do, and weren't capable of judging anyone due to their alien makeups. The times spent in their home were the best parts of his days. Which reminded him he needed to make a mental list of things he wanted to discus with them later this day.

As the final bell of the day sounded, and Mickey went to find his bus, a stomach-turning, yet perhaps fitting end to his day occurred. Parked across the street from the school, at the end of the line of buses, sat the clearly marked vehicle with the City of Cypress Springs Police insignia stenciled on the front door. Behind the wheel sat the chief, wearing dark glasses, which did nothing to hide the fact he was staring directly at him. Increasing his walking pace, Mickey soon broke into a run, searching for the sanctuary of the yellow bus.

CHAPTER TWENTY-NINE

Mickey made what amounted to a pit stop at home to rid himself of his book bag, and to check up on Jan. He managed to peek his head in to her room before his mother shushed him away. All he could see before being banished was a lump of bedding covering an otherwise inert, trim body.

"How's she doing," he asked.

"Not much better, I'm afraid. Might be a long weekend if she doesn't show improvement overnight." She looked closely at the boy. "You not feeling anything so far?"

"No, ma'am. I feel fine. Kristi's still missing, so I guess she's got it pretty bad."

Jessica put a hand to his forehead for good measure, her nod indicating he was in the clear so far.

Later, just before approaching the door at 710 Louisiana Street, Mickey was struck by the thought that the "Jones" family might be concerned about having been exposed to whatever Jan had. He needed to inform them of any potential illnesses they may have been in the company of.

Once inside, Heidi, who was alone in the living room when greeting him, assured him there was no cause for concern.

"For some reason," she explained, "we appear to be immune to most contagions that affect humans. We were surrounded by various strains of what you call flu viruses while in Michigan, but managed to avoid them ourselves. Possibly due to the design of our built-in immune systems. Our planet has its own diseases, but yours don't cause us concern."

Mickey nodded as if he understood, relieved at her answer. But it also produced another question. "You said the icy winters in Michigan were bad for you. Does it ever get cold on your planet?"

"Actually, it does not have those extremes. The primary reason is that our sun, although somewhat smaller than yours, is much closer to us. In addition, our planet's revolution around our energy star has a much shorter duration than Earth does around your sun, which means we are not faced with cooler conditions for long periods."

He understood part of what she had said, but was lost when it came to the revolution part. Not wanting to show his lack of knowledge, he pressed on.

"I guess our colder weather now in Cypress Springs isn't welcome to you."

"We manage to stay inside as much as possible. As we explained to you and Jan, except for my going to school and back, we rarely leave the house. Plus, this is nowhere close to those conditions in the upper part of Michigan."

As they took their obligatory seats on the sofa, she added, "By the way, I am sorry to learn Jan is ill." After pausing, she asked, "Is she enjoying our sessions?"

"Oh yeah. For sure. It's all she's talked about when we're alone, especially after yesterday afternoon. She's missing not being here today."

"Good. I thought she might. She appears intelligent, and asks such interesting, insightful questions."

Assuming insightful was another word for curious, Mickey fumbled with his hands for a second, stalling, while building up his nerve for something that may be too bold to ask for. "Speaking of questions, I have one that may be hard to answer, or agree to."

"Go ahead."

He looked toward the rear of the house, beyond the kitchen. "Is there any chance you can tell me what goes on back there? Will I ever get to see any of it?"

Heidi hesitated for only a second or two. "Actually, we have discussed that. But before I answer, let me confer with the others."

With that, she quickly rose, gliding from the room into the kitchen area, then supposedly into the rooms behind. As he waited with an apprehension similar to the feeling he had when awaiting a test score, Mickey tried to relax. He figured the worst that could happen is they would decide "no". But the possibilities were so great, it was hard to remove the tension from his body.

Instead of Heidi reappearing alone a few minutes later, she walked in behind Rhett and Jed, who seated themselves in the usual chairs. It was Rhett who spoke, leaning his flesh-colored arms on his knees.

"We have agreed the time will come when you should see what we have stored in a rear area. In fact, you need to observe it to fully understand us, and our capabilities. However, we feel certain it will be more beneficial for you to see something else first, since it is connected to what you will discover back there." He pointed with a turn of his head that almost dis-lodged his wig again.

"Okay." He was excited, yet anxious about this new twist.

"How would you like to visit our transport vehicle?"

Barely able to reply at first, the stunned boy managed to choke out a "Yes, sir. I mean, Rhett. More than anything." His chest was beginning to heave, almost as much as that first amazing afternoon when Heidi and Blanche stopped him from leaving. The offer was better than his original request.

"Very well," Rhett said. "It will require some planning, both by you and ourselves. I will need to rent a vehicle to deliver us to the site, so you simply must tell us when you can spare an hour or two after school so that I can have it ready."

"How about tomorrow?" he quickly offered.

"Well," Heidi said, "I had wanted to have Jan accompany us. Do you think she will be well by tomorrow?"

"Oh, that's right. I want her to see it too. But, seeing her condition today, I doubt she'll be up to it." His hopes were immediately dashed as suddenly as they had materialized.

"On second thought," Jed spoke up. "It might be best if Mickey goes alone this first time. I think Heidi and you, Rhett, should go also. And, of course, myself. Blanche can stay behind and monitor things from here. The vehicle you rent may only carry four comfortably. Jan can go later, if she desires."

"I can guarantee you she will," Mickey said quickly.

"Then, it is settled," Jed said, adjusting his comical looking hat. "We'll go about this time tomorrow, and have you home by six at the latest. Just get word to Heidi if you need to postpone for any reason."

He answered, "Okay", knowing full well there was nothing on God's green Earth that would require him to miss the reveal of a lifetime tomorrow promised.

CHAPTER THIRTY

Friday proved to be the longest day of Mickey's life, the hours dragging by with the promise of another disclosure by his new alien friends waiting; this one more exciting than all the rest. The only downer was the gathering of what could be storm clouds arriving by mid-afternoon. He didn't know if that would alter plans or not, but there was nothing he could do about it if it did. He would show up at Heidi's house and hope for the best.

Something else that might affect his plans for later in the day was Kristi's return to school. She looked a little on the pale side, and her blue eyes weren't as sparkly as usual. Although glad to see her recovery, he couldn't see how he could ever sit through one of their get-togethers after school without giving away the fact there was somewhere else he wanted to be. Fortunately, she solved that dilemma for him. Approaching him on his way to the lunch room, she admitted she may have come back too soon; she was having a sinking spell, and was headed to the office to call her mom to come pick her up.

"Gosh, I'm sorry, Kristi," he commiserated, telling her to get some rest over the weekend. "Maybe we can meet Monday," he said, sounding as hopeful as possible. He was a free man, or at least, a free boy. But at the same time, having thoughts like that had him thinking about some of what

Karla had told him a couple of weeks ago, the afternoon she explained what he took to be some of the facts of life. Maybe the attraction was starting to weaken, and his interests were changing; or maybe hers. He would have to find time to think about that.

Jan had broken her fever sometime during Thursday night, and was feeling better, although without much energy. She had insisted on Jessica's going to work Friday morning.

"Look, Mom. I'm sure Mr. Montgomery needs you more than I do today," she had argued. "I'll be fine. Don't forget I'll be fourteen in only a month, so I can make it by myself."

Overhearing that discussion, Mickey wrestled all day over whether to tell his sister about the offer made to allow him to see the transport device that had brought the Jones' to Earth. He was afraid she would insist she was well enough to go—in fact would demand to go. He could reason with her, explaining the stress of getting dressed, plus the chill of the early evening air would cause a relapse. Then there was the truth of what Jed had pointed out; there would only be room for four travelers, although that may-or-may-not be correct. Neither of those arguments would likely be enough to talk the strong-willed girl down.

Of course, he could tell her nothing, and deal with her wrath when admitting things after the fact. But if there was one sure way to bring out her dark side once again, that would surely be it. Having changed his mind five times during the day, he decided in the end to judge her mood and condition when arriving home.

She was wearing the quilted robe Louise Knox had knitted her earlier in the year, resting in the den recliner, watching tv. Ankle high socks were fully visible peeking out of the bottom of the robe; the girl had grown so rapidly the past few months, it would cease to fit her soon.

"Hi, Mick," she greeted meekly. "How was school?"

He sized her up before answering. Her eyes were puffy, and a few shreds of Kleenex were spread across her lap, indicating she was still not quite there. His decision was pretty much made for him.

"Fine. How you feeling?"

"Ehh, better, but still not to shiny." She sneezed loudly into the cleanest tissue she could reach.

So, he told her about what was going to happen right away. She did sit up straighter, as if ready to object, but her recognition of her weakness put her back down. He hurriedly added that she would be joining him as soon as she was well, which seemed to help.

"But I want to know everything you see. Every single detail of the . . . whatever it is." She considered what else to ask for a moment. "Are you going to go inside it, do you think?"

He shrugged. "No idea. Just have to see how much they'll let me see."

Getting rid of his bag, he made a bathroom stop, then headed back to the front of the house. "Can I get you anything before I go?"

She requested a bottle of Sprite, which he quickly produced for her. As he turned to leave, she stopped him.

"Hope you enjoy whatever you get to see today, Mick. You know, you are one lucky guy. Have you thought about that?"

"Yes I have. But we are both lucky, Jan. We're doing this together."

"Yeah," she replied, with a wry smile. "except for today."

Thinking they were getting a little too mushy and sentimental, even though the words they spoke were the truth, Mickey quickly left, walking out under crummy gray skies.

The rental was a white Ford Fiesta hatchback, which was definitely built for four passengers. Any more and the drive could get uncomfortable. It sat in the open carport at Heidi's house, appearing to Mickey to be too plain and unremarkable for the mission it had in front of it. He often let his imagination put inanimate objects into living forms. It made life more interesting and dramatic for him.

Entering the front door with Heidi, as usual, opening it seconds before he arrived, he quickly asked if the weather was a problem.

"Not at all," answered Rhett, who was also in the room, along with Jed. "In fact, it may help provide cover for us. We certainly don't desire to draw any attention to our journey."

The word "journey" made him wonder if they were going some distance. As it turned out, he wouldn't know for certain how far they had traveled when all was said and done, because as the four of them, Jed, Rhett, Heidi, and himself, climbed into the car, Rhett said, "We hope you will not take offense at this, Mickey, but we must ask you to wear a blindfold." He produced one from a pocket on his shirt.

"It is for your benefit as well as ours," Heidi quickly added, noticing his hesitance in reaching for the black cloth item. "We do trust you completely, Mickey. But it is conceivable someone at some point may attempt to force you to divulge where we have kept our transport. If you aren't aware of the location yourself, then that cannot happen."

"Sure. No problem." He probably sounded more confident than he actually felt. Given the situation, for only the second time over the past few weeks, he was questioning things. There was, he figured, the slightest chance they might be abducting him, for whatever reason, he had no clue. But with Blanche still in the house, and Jan also knowing their story, he dismissed the paranoid thought.

"Also," Rhett said, "it may also help if you can lower yourself beneath the window level until we have traveled for a while. We had rather not give the good people of Cypress Springs the impression you are being taken against your will."

Speaking of drama; this was getting heavy. "Okay," he said, letting Heidi, who was in the rear seat with him, tie the blindfold tightly in place.

"Well then," Rhett announced, "we are off." Behind the wheel, he backed onto Louisiana, with Jed riding shotgun, Heidi in the back, and with their guest passenger scrunched down as far as he could go comfortably.

Mickey had never been good at guessing at the length of time passing, whether in school, church, or anywhere else. But he *could* figure out directions of movement, even, he soon discovered, while blindfolded. Squatting on the rear floorboard, he paid attention each time his body shifted as the vehicle made turns. Starting out, there was only a single right turn, which he could easily figure out was the compact car turning off Louisiana on to Sabine, passing right in front of his own house. Then, it was straight for about five minutes or so, followed by a series of left turns, causing his lightweight body to shift to the right at each turn. This lasted for roughly five minutes, giving him time to figure out they were circling the town square in an attempt to confuse him. It was working. He had lost count of the turns.

Finally, the car straightened out, then after one last turn to the right a few minutes later, the Ford opened up, increasing it's speed on another, longer straight-away. His knees were starting to cramp slightly, and his elbows were sore from bracing himself against the back of the front passenger seat. But at least, the turns had ceased.

"How are you doing, Mickey?" Rhett asked from the front.

"I'm okay so far."

"Good. You will be able to remove the blindfold and sit up in a few minutes, so try and make yourself comfortable."

Contrary to Rhett's promise, the car continued for what seemed to be about fifteen minutes. The three aliens, which Mickey was now able to consider them as, spoke no other English as they traveled, but did make some sort of sounds. They must have been speaking in their native language, which seemed to consist of mostly zzzz's and rrrr's, with a few other sounds, which may have been consonants, included. Evidently, they were reluctant to speak in words he might understand, words that might give the destination away.

When the car finally slowed and made another turn, it was to the right once more. The smoother highway they had been on gave way to a rougher roadbed, which seemed to be paved, but poorly maintained. He was forced to press his arms against the seat ahead to keep from being thrown around. As rough as that was, it was nothing compared to the conditions following the next turn, which was about ten minutes later, although it was so unsettling it could've been much shorter or longer. It was only a wild guess, but he figured they had to be at least fifteen miles from town, but in which direction, he had no clue, having lost his concentration.

A few minutes later, the car slowed, then stopped. There was more alien chatter between Jed and Rhett, then a similar voice seemed to come over a speaker of some sort. Then, the car started up again, but still at a slow pace.

"Okay, Mickey," Jed spoke. "You can sit up, but keep the cloth over your eyes for only a few minutes more."

Another few minutes passed before the small car came to a halt once more, this time proving to be a permanent stop.

"Okay," said Rhett. "Everybody out. You may remove the hood, Mickey."

Stepping from the car while removing the blindfold, Mickey took a minute to flex his knee joints, getting the cramps out. He rubbed his lower arms that had been pinned against the front seat. He took a quick look toward the sky, hoping to find the sun to get his bearings, but the cloud cover hid that clue. Looking around, he saw the Fiesta had come to a stop in a small clearing, surrounded by trees, the opening only large enough for the vehicle with only a few yards to spare.

Rhett and Jed had moved to the edge of the clearing, now looking back at him.

"Well, what do you think, Mickey?" Jed asked.

Confused by the question, he shrugged his shoulders. "About what?"

"Why, our transport, of course."

Mickey took a 360 degree scan of the surrounding trees, seeing nothing resembling a space pod, or whatever it was. "But, where is it?"

Jed allowed himself a thin smile, unusual for the aliens. “Very good. I did not mean to trick you, Mickey. But you were a perfect test of how well we have hidden our module.”

Jed turned to look in the direction of the car, holding a hand out. There was something he was holding in that hand, about the size of the phone they had let Mickey use earlier. Activating a button, the underbrush approximately twenty feet away, on the opposite side of the Fiesta, began to rise, accompanied by a faint mechanical sound.

As Mickey moved to stand next to Jed, the underbrush completely fell away, revealing a gun-metal gray curved surface that seemed to be floating above ground. It appeared to have a height of an adult—perhaps six feet. Although he had adopted the belief that nothing these alien beings did could surprise him any more, he was struck dumb by what resembled a turtle-like shell rising from the Earth. Jed approached the shell, with Mickey following on legs he could no longer feel.

“You see, Mickey,” Jed explained as they approached. “We knew if you could not detect our hiding place, hopefully no one else could either. We apologize for using you as a test.”

Mickey wasn’t listening to a thing the man under the floppy hat was saying. He was transfixed by what was in front of him. Drawing to within a few feet, he could now identify a second shell, inverted in comparison to the first, lying in a hollowed-out bed of more underbrush. He started to reach his hand out to touch the upper surface, when Jed stopped him.

“You should not touch it yet, Mickey. It possesses just enough energy to give a small shock. Not to injure, but to warn others away.” Jed then reached underneath the bottom rim of the shell, touching something

Mickey couldn't see. The shell layer flashed a dim light briefly. "I have deactivated the charge. Go ahead." He nodded for the boy to do as he wished.

As if touching a hot plate, Mickey gingerly moved his hand back and forth a couple of times, then reached to lay his palm on the surface of the device that had come from a world so far away, no one on planet Earth but he and Jan knew it existed. Feeling the overwhelming adrenalin rushing through him, he rubbed his hand across the surface, then turned back to see satisfied looks on the other faces, especially Heidi's. A gust of wind suddenly rattled through the trees; Rhett and Heidi had to hold their hair pieces in place with a hand. Jed pulled his floppy hat lower.

"I don't know what to say," he blurted out. "Or how to act. This is totally unbelievable." Getting more than a little emotional, he looked back at the transport in wonder. "Do these two parts fit together?" he asked, looking at the top and bottom.

"Yes, exactly," Jed answered. "It requires a few steps, but within a very short time, it would be equipped and ready to lift off."

Completely awestruck by the amazing experience, Mickey took a few minutes to circle the transport, working his way through the brush. It was roughly the size of a large room, maybe fifteen feet across. Viewing it as it sat there, in its inert state, he found it hard to believe that this contraption was the reason for those bright lights flashing across the Texas skies the previous spring. And *this* was where it had landed. It was almost too much to deal with. He was already anticipating trying to explain what he was seeing to his sister. He wished she were here with him so that he wouldn't have to.

After Jed and Rhett spent time checking out the perimeter surrounding the hiding place, and for some reason looking up into taller trees nearby, Jed announced it was time to go. He leaned in to turn the electric charge back on, and the lights flashed briefly again. He then hit the same button in his hand-held device, and the bottom half of the pod lowered itself back into its mooring place, then the top portion was lowered into position. All four of them, Mickey included, spent a few minutes tossing the underbrush back into place. Finally, after admiring their handiwork, within a half hour of arriving the group climbed back into the Fiesta, and drove back into town.

CHAPTER THIRTY- ONE

Before the small car was turned around for the drive back into Cypress Springs, the blindfold was re-applied. Thankfully, Mickey was allowed to sit on the back seat, at least until they neared the outskirts. Then, it would be back on the floor. Even though he couldn't see, he could still talk, and he had a new salvo of questions.

"I could see how well hidden it was, but aren't you afraid somebody, like local property owners, mighty stumble across it?"

"Yes," Rhett said. "That is possible. But we are able to observe the activity in the area on a daily basis. As of yet, no one has come near."

"Observe? How? You can't come out here every day?"

"You are correct, Mickey. We chose not to point them out to you, but we have installed several rather small video cameras in the area so that we can monitor what was occurring in the vicinity. Which is why we came to a temporary stop before actually going into the final area. We were testing to make sure that particular camera in a tree nearby was functioning properly. By the way, the road into the tree stand where our transport is hidden ends shortly beyond that location, so we feel safe in that no one will approach from that direction."

Mickey looked in the direction of the man speaking, as if he could see him. "But how could you tell the camera was working just by looking at it?"

"Another excellent question, Mickey. All of the cameras are transmitting images to several screens in our home. Blanche was observing us as we approached. She would then communicate to us through ear plugs Jed and I are wearing. So small, you likely have not noticed them. We check those screens several times daily."

"So, that's what's in that back room? Television screens?"

"Yes," Jed answered. "Among other things. However, they are not the kinds of television screens you are accustomed to. They are only closed circuit monitors."

Closed circuits was something he had heard of, but had no idea how they worked. The two aliens were describing things that sounded incredible. He wasn't sure he believed all of it, but thought it wise to act as if he did. He didn't understand most of it, but it did make sense, if it existed. He moved on to another question.

"I didn't see any chairs or anything in either half of the craft. Where did y'all sit? And I didn't see any motors or engines, or whatever it takes to make it fly."

He could tell Jed was looking at him as he answered. "All of the interior fixtures collapse into the bottom section. Seating, heating and cooling devices, as well as all of the navigational and communication equipment is located there. However, we must also bring other pieces of equipment from home if we are to launch the craft. I know it all sounds rather involved, but remember, we have had many centuries to refine our technology to allow scaled down versions of all that is required for us to transport us quickly and efficiently."

Mickey was once again blown away by what he was learning. At the same time, he couldn't believe things were as simple as he was being lead to believe. His thoughts were interrupted by Rhett, who he could tell was also looking at him.

"You'll see once you are inside the craft how it all works."

"Okay." *Wait! What did he just say?* He leaned with his hands draped over the rear of the front seat, wanting clarification. "Did you just say I'm going to be able to get inside?"

Mickey heard Heidi, sitting next to him, actually release a chuckle, or at least something sounding like one.

Rhett, again driving, evidently took his eyes off the road long enough to answer. "Yes, Mickey. We have been discussing that very thing. It will be useful to our observations, and for your own knowledge, if you were to take a short excursion flight with us. Jan too, of course."

Mickey threw himself back against the rear seat. He wanted to scream out loud, but knew that would be too uncool. All he could think to say was, "When?"

"Soon," Jed answered. "Keep us advised of Jan's progress. Plus, there are a couple of other things to consider. Allow us time to look at a few factors. It will take planning on all of our parts."

By that time, they had reached the city limits, and Mickey had to get on the floor once more, until they were safely within a few blocks of their neighborhood. Able to sit up and remove the hood again when reaching the house, Mickey was anxious to head home and try to describe to Jan what he had seen, and more importantly, what lay ahead for them. But before he could leave, Rhett went over a few points to keep in mind.

"According to your calendar, the daylight hours will shorten in slightly over a week, on October 28, with the end of Daylight Saving Time."

"Oh. Is that bad?"

"In fact, it is good. It will allow us to travel in the gathering dark for most of the trip we have in mind. For planning purposes, it would be ideal to be in position at the site of the transport to depart shortly after five p. m. The sun will have set soon after, and we should be able to return at about seven p. m. That means you and Jan must plan to be here, at our home, no later than four-fifteen. Make sure she is aware she must be blindfolded also."

"I will. She'll understand." It was if they were planning something no more exciting than a school field trip. He'd forgotten for a moment what they would be traveling in. "Where will you take us?"

"We have an idea, but leave that to us." Rhett was thinking about something else. "Do you think your mother will be concerned if it is after seven in the evening before you return?"

Mickey thought about that. "I don't think so, as long as Jan and I can come up with a good reason why we'd be late. Let me talk to Jan, and I'll let you know."

As he started to leave, he turned back. "Guess I'd better spend time with her this weekend, so she doesn't get suspicious."

"Of course. Whatever is best for you."

"Oh, and thank you for letting me see that tonight. It was all . . . amazing."

He left the others as they prepared to enter their house, beginning his walk to his own. He had a feeling similar to walking out of an exciting movie, reliving the scenes in his memory. But, this . . . this was the real deal, not something filmed on a movie lot somewhere. It was easy to think it had all been a dream—an unbelievable dream.

He was so lost in his thoughts, he hardly noticed a form stepping from behind a tree bordering Heidi's front yard. Dusk had settled just enough to help hide the figure. Almost before realizing someone was there, a deep voice interrupted Mickey's progress.

"Good afternoon, Mickey," Winston Dunaway said. "Or, should I say, good evening. Shouldn't you be home by now?"

Too terrified to answer, Mickey came to a halt, sinking back. Taking a frightened glance around, he saw it was just the two of them, standing only feet apart. As he had that night in his living room, the man seemed ten feet tall, towering over him. The sea of white in his eyes stood out in the gathering gloom of early evening. Mickey hadn't noticed the chill in the air until now, which was causing him to shudder. The man seemed to be waiting for a reply of some kind, but the boy still couldn't speak. *This was it,* he figured. He was going to meet his doom, all alone with no witnesses.

Dunaway filled the void, as he also took a look around. "You never know, Mickey, what might happen when you're out by yourself. You just don't know who's out here."

"Yes, sir," he managed to utter in a high, squeaky voice.

That produced a satisfied smile on the man's hard face. He stepped forward a bit, more like a shuffling movement. "Another thing you ought to keep in mind is, some people, especially kids like you, see things they don't really understand. And sometimes what they see can get themselves, and a lot of other folks in a mess of trouble." He moved a little closer, his large boots coming within a couple of feet of Mickey's sneakers.

"Do you understand what I'm getting at, Mickey? I think it's best to just forget about things that look strange to you. You're too young to get the whole picture. Can't do any good to hang on to those things."

"Yes, sir. I think I understand," the boy answered through chattering teeth.

"Well, that's good. Yes, sir, that's good." Dunaway started to back up, then paused. "Now, I think you ought to get on home to that pretty little mother of yours. Sure she's worried about you."

His last words struck Mickey like a punch to his frigid gut. He was savvy enough to recognize a veiled threat.

"You leave my mom alone!" he said with more force than he thought he had. He was glaring up at the giant, who appeared surprised by the outburst.

The big man leaned in. "Now, see here, boy. Don't you know who I am? I can cause you a heap of trouble, whether you keep your smart mouth shut or not. You better . . ."

That's as far as he got, his words now frozen in his mouth, just like his huge body that was seemingly frozen, locked in place. His weird eyes seemed to widen ever so slightly, and the faintest odor of something burning struck Mickey's nose. He hadn't seen the scaly hand reaching through the darkening air, but he saw it withdrawing from the chief's back.

Stepping back a few more steps, Mickey understood with a degree of horror what had happened. It was Rhett who had appeared, laying a charge of his hidden energy on the man with a mere touch. He now approached, standing to the side of the huge, inert lawman. Mickey's fear was that the man would be even madder now, coming for him with a new vengeance.

"Go home, Mickey," Rhett said calmly. "He will not remember what has occurred."

"Are you sure?"

"Oh, he may recall coming here, but beyond that, nothing. As Heidi explained, we can activate any level of our defensive ability as we chose. Unfortunately, I found it necessary to use a higher level than she used on your friend, and much higher than the night she and Blanche convinced you to stay."

Mickey found it hard to believe what Rhett was saying. Dunaway's eyes were wide open, staring right at him. "Are you certain he can't hear us now?"

"Absolutely. However, whatever it is that you may know that threatens him, you may want to keep your distance for a while."

"You should tell that to him."

A rare grin crept onto Rhett's plain face. "If only I could, Mickey. Now go. His condition will last for approximately two more minutes."

Mickey needed no further encouragement. He made it home in record time, never stopping to look back.

Since he had been leaning down in the direction of the boy, not that he could recall doing so, when Dunaway broke free of his unconscious state, his momentum caused him to topple forward, barely catching himself with

his hands before hitting the ground. He quickly pushed himself to his feet, taking a sweep of the area to try and determine where he was, and why he was there. No one else was anywhere to be seen. Where the heck was he?

Slowly, pieces of his existence started coming back, like dominoes falling one by one. He could vaguely recall seeing the boy—the Foreman boy. But why had he come looking for him here, wherever here was? It would be several minutes later, after had he managed to find his car and climb behind the wheel, before it came to him that he could remember the night he had surprised the boy by showing up at his mother's house. While there waiting, the woman had explained that her son had been spending time lately at the house of a girl name Jones. She lived on Louisiana Street, the street where he had just ended up in someone's yard. He now recalled seeing the boy come out of that house before stopping him, but beyond that—nothing.

The only thought that seemed to make any sense in his addled condition was that he needed to find out about the girl in that house, and maybe her family while he was at it. Whatever had caused him to lose those minutes, he was pretty sure the boy couldnt've done that himself. Impossible!

CHAPTER THIRTY-TWO

For the rest of the night and into Saturday morning, no good time presented itself to talk to Jan about any of his Friday afternoon adventures, both good and bad. Another pizza delivery was on the menu Friday night, with Jan assuring her mother her appetite had returned enough to put solid food in her stomach, and keep it there. The Foreman family ate at the kitchen table in mostly subdued silence, each family member with their own reasons for their moods. Mickey was busy assimilating what his afternoon had been; Jan still on the fading edge of her bug; and Jessie simply winding down from a pressure-packed week at work. When there was even the briefest of opportunities presenting itself, Mickey would shoot some sort of non-verbal sign to his sister, who seemed to recognize the excitement on his face. A thumbs-up or a Cheshire cat grin told her quite a bit.

Finally, an early Saturday afternoon hair salon appointment, an infrequent occasion due to budget constraints, had Jessica out of the house for a couple of hours. Since the co-conspirators had shared their earlier secrets in Jan's room, that was the chosen spot for the latest news. The good parts came first.

"I wish you could've seen it, Jan," Mickey said, after describing the scene at the pod site, and the gadgets Jed had demonstrated.

"And you have no idea where the place is?"

"Not really. I could follow things at first, just by listening and feeling my body shift at the turns. But we made so many turns before leaving town, I had no idea which highway we left town on. We ended up so far out, it could've been anywhere."

Pausing a moment for the proper effect, Mickey moved on to the big news. When he told her, she jumped from the bed in spite of her weak condition, pacing the room excitedly, while using a Kleenex to wipe her nose that continued to run.

"Are you kidding me, Mick? You better not be." She pointed a finger at him.

"No, Jan. This is going to happen, but you have to get well first." He then went through the things Rhett had pointed out, about waiting until the days were shorter, and the chance they may be getting home late from wherever they were going.

"Do you think Mom will buy any excuse we can come up with for being late one night?" he asked.

"Let me think about that. First thing that comes to mind is we can say we're having supper at the Jones' house. But that may seem weird to her, since she's never even met them. I'll be able to think clearer when I finally get rid of this crud." She took another swipe at her nose.

"Right."

Then it was time for the not-so-good news. He described the encounter with the police chief, and his rescue by Rhett.

"I was about to tell him I'd keep my mouth shut, then he mentioned

Mom. I got a little hot then. I hope Rhett was right about the guy not remembering everything that happened."

"Gosh, Mick. The man's getting desperate. Maybe we ought to talk to somebody about him."

"Who? He's the Chief of Police, Jan. There's nobody over him.".

"There's always the sheriff, whoever that is. He's the man in charge of the whole county, so he's gotta be over Dunaway."

"But, it would just be my word against the Chief's. No way that'll work."

"Guess you're right about that. But, Mick. Even if Rhett's right, and the creep doesn't recall what happened. He'll still worry about you whether you keep quiet or not."

She was right, of course. Which meant one more thing to lose sleep over.

Saturday night was spent by Mickey at Benny Payne's house. Just as he and Jan had completed their secret meeting, Benny had called, inviting him to an early family spaghetti dinner and to watch a movie on their VCR. Mr. Payne was an officer at the largest local bank in Cypress Springs, which allowed the Payne family the where-with-all to purchase the latest trinkets, including the cassette recorder, which continued to amaze Mickey.

"The movie is *The Invasion of the Body Snatchers*, Benny said. It's not the stuff we usually like, but sounds kinda spooky," Payne had said.

That didn't matter to Mickey. It was a good way to get out of the house and maybe get his mind off other things. Still leery of the chief lurking somewhere in the dark, he surprised his mother by asking her to drive him to Benny's. She was indeed shocked, but happy to do so.

"The Paynes' will drive me home," he said as he climbed from the Toyota at Benny's house.

The Payne home was a comfortable two-story bricked job, one of the nicest in town. The plush decor and fancy-looking, dark wood furniture was some- what intimidating to Mickey. He was always afraid to put his soda glass down anywhere without checking to get approval from Mrs. Payne, who would wave off his concerns.

"Anywhere is fine, Mickey."

Adults, with the exception of Chief Dunaway recently, were usually friendly toward him, which he chalked up to pity due to his less than normal family situation. He took advantage of being in the fashionable home, checking out the furnishings, making mental notes about what he planned for himself at some point in his future.

Soon after Mickey's arrival, he was surprised, and more than a little pleased, to see Trey walk through the front door. Benny had taken the lead in trying to put the small gang back together again, inviting the rusty-haired boy as well. Roby was missing; no weekend nights ever found him free from Senor Pepe's. Things were awkward for the first few minutes between Mickey and Fulghum, but by the time the spaghetti had been consumed by the two, along with Benny, his parents, and his two older sisters, the tide seemed to be turning. The time spent watching the movie, which was definitely on the creepy side, with a surprising, disturbing ending, reminding Mickey of his encounter with the Chief, completed the rekindling process of the friendship. So, at least that one part of his life was trending in the right direction.

Eleven year old boys rarely voiced their inner feelings over developments in their lives, least of all to other boys. The same went in regard to apologies. This was one time Mickey considered for just a few seconds breaking that code, again recognizing his role in Trey's treachery. The scars were beginning to vanish from Fulghum's face, but Foreman couldn't escape the fact that his friend was the only person who had been injured by the collective secret—at least so far. In the end, he decided not to mention anything about the events of the past few weeks, hoping against hope it would all soon be forgotten.

Although the menacing image of Chief Dunaway's threats popped into his head several times during the night, the time spent with his friends had served as a salve for recent unsettling events in his young life. He did his best to concentrate on the coming adventure with Heidi and her alien family. But even that thought caused him grief. Sooner or later his friends would come to realize he had excluded them from the adventure of all adventures—a chance to learn first hand about life beyond Earth's boundaries. Resentment would be expected, and no doubt justified. He wished it could be otherwise. On the other hand, there was no one he had rather share the experience with than Jan. They would simply have to accept it. If they were truly friends, they would.

CHAPTER THIRTY-THREE

Monday morning found Chief Winston Dunaway wrestling with what he could do, if anything, about the occurrence of his Friday afternoon black out—for the lack of a better term. It was more a question of what he *should do*, since maintaining a low profile might just be the way to go. Making too big a deal in too loud a voice could even be worse than letting things go. It was possible he had gotten his point across to the boy that he should keep his nosy mouth shut. But since most of the encounter was blank to him, he wasn't sure about that.

The big worry that kept crossing through his mind had to do with whoever had interfered with his talk with the Foreman boy must also know the reason for the talk. The help the boy received had to come from that person, and that person must reside in the Jones' household. He knew the name again thanks to the Foreman woman's mention of it. So he was making it his unofficial duty to find out what he could about the family located at 710 Louisiana Street. The simplest and quickest thing was to scan the phone book, which produced no Jones' on Louisiana Street. That didn't completely surprise him, since things weren't normally that easy. He would have to dig deeper, hopefully without appearing to dig at all. Using his position of authority, he made a discreet call to the voter records department in the nearby county administration building. No one with that name at that address on their rolls either.

A call to the property tax department in the tax assessor's office produced a mild surprise. The address in question listed as tax payer an estate; the estate of Opal Frederick, etal, which indicated to Dunaway that the house must be a rental. The trustee of the trust established by the estate was listed as an attorney he was familiar with, and on a first-name basis. Placing a call to Tom Jagger, the attorney in question, he left a call back message with the man's secretary. As he waited for the return call, he ran through any other avenues available to him, short of knocking on the Jones' door and interrogating whoever was inside, which he knew would be foolish.

Forty-five minutes later, Tom Jagger returned the call, verifying the former home of Opal Fredericks was now under lease. There was a sister in Oklahoma who, as the only heir, and beneficiary of the trust, had decided to hang on to the property, especially after initial offers to purchase weren't attractive to her. High interest rates were keeping buyers at bay.

"Mind if I ask who handles the rental?" Dunaway asked in a level tone.

Following a short pause, the lawyer evidently chose to avoid any challenge to a need-to-know he could offer up, giving the name of a local realtor who had handled the deal for a modest monthly fee taken from the rental payments.

The next call, hopefully the last, was to the realtor, who seemed more than happy to help out the Chief of Police.

"Yes, Chief, the property has a one-year lease on it, renewable with two months notice by either party at the end of the term. Handled it myself. Got a good deal for the estate, if I do say so myself."

And for yourself, too, Dunaway thought to himself as he sat through the man's self-congratulations and legalese talk.

"What can you tell me about the tenants? Know anything about them?"

Dunaway could hear the sounds of a file cabinet being opened, accompanied by the man's grunting as he reached to extract a file. "Let's see here. Oh, yeah. Supposedly a family of four, but I only saw two of them. A Rhett and Blanche Jones, husband and wife, near as I can recall."

"What about the other two?"

"Mmm. Oh yeah. I see in my notes there was also a daughter, and a parent of one of the two—didn't make note of which."

Dunaway thought for a second or two. "Any mention of where they moved from?"

The rustle of pages being turned followed. "Not in my notes, but seems I recall they said something about coming down from the north. Too cold for them up there," he added with a chuckle.

"I assume you did some sort of credit check on 'em, making sure they could handle the rent."

"Well, ordinarily I would. But seeing how they paid the full year in advance, didn't think it was necessary."

"The full year? How much was that?"

"It was exactly six thousand, six hundred dollars, which is five-fifty per. For that deal, the sister and I agreed to do all the maintenance and upkeep."

Dunaway's police antennae was straight up now. "What did they list as occupations?"

"Why, nothing at all. I got the impression they were pretty well off. Must not need to work. In fact, they paid in cash. Another reason for no credit check"

That antennae was now beeping louder. "Cash? Why the heck didn't they purchase instead of rent, since they seem so flush with money?"

"You got me, Chief. Some people like to try out an area first, then buy later if they like the place. Plus, they were so anxious to do the deal, I didn't stand in their way."

"Did they say they have any ties to the area? Any relatives, friends?"

"Not that I recall, and nothing in my notes."

"And you're sure they didn't say where they were from?"

"Can't really say, Chief. But my guess it was maybe Wisconsin, Illinois, or even Pennsylvania."

Dunaway shifted in his chair, beginning to take notes on the conversation. "What makes you say that?"

"I do recall them saying the winds off the lakes up there made winters unbearable. So, sounds like the Great Lakes to me."

"Makes sense." Dunaway started to end the call, then thought of something. "What about social security numbers? I assume you took those down."

The realtor cleared his throat, indicating discomfort. "Well, there again, Chief. Didn't think it was necessary, being how it was handled. And they were very nice folks. Real polite." Then, after a pause, "Mind if I ask why you're bringing these people up?"

He was ready for that one. “Oh, some neighbors just wanted to find out about them. Can’t say who. You know how it is.”

“Sure. I see.”

Thanking the man, Dunaway hung up, rocking back in his chair to think things through. It was beginning to seem the objects of his curiosity were either running, or hiding, which are really the same thing, in his lawman’s view. Two scenarios came immediately to mind: the relatively new, so-called Witness Protection Program, or organized crime possibly drug related. With a lot of cash money on hand, drugs made a lot of sense. A third reason soon entered his mind; illegal aliens, possibly from Canada, since they had come from up north.

The Witness Protection deal was completely off the table. From what he had read and heard, those displaced individuals were beyond the reach of local law enforcement, similar to diplomatic immunity, wrapped up in a cocoon by the feds. He wasn’t about to ruffle those feathers. That left the Drug Enforcement guys, and the Immigration and Naturalization Agency, the enforcement arm of tracking down and dealing with people in the country illegally. He knew bringing the DEA into the picture could easily open a whole new can of worms, a situation he wanted no part of. Something he recalled from a conversation with one of his officers suddenly came back. The man in question had recently been to the Law Enforcement Training Center in Dallas, along with others from around the state. One of his fellow trainees had gone to work for the INS after completing the course, according to his officer.

Picking up the intercom, he waited for his receptionist to come on the line.

"Shirley, is Officer Sweat in the building?"

"Pretty sure, Chief. I can see his vehicle through the side window."

"Have him come to my office. I need to have a talk with him."

CHAPTER THIRTY-FOUR

Following the Saturday spaghetti and movie night at Benny Payne's house, Mickey began to think that might provide an answer to how to arrange to be out late when the big adventure finally arrived. It seemed like the perfect story for he and Jan to give their mother that would seem logical. But all kinds of potential snags quickly came to mind. Jessica Foreman and Mrs. Payne weren't that close, but you never knew who would bump into who in a small town, in the grocery store, or drug store, or anywhere. Their cover could easily be blown. Then there would have to be a reason for his sister to accompany him to Benny's. Jan hardly knew the older Payne girls, so that wouldn't make sense. In the end, he decided to rule out that ruse.

Later during the week, the problem appeared to solve itself, thanks to Jessica's job. Early in November, an annual contractor's conference, with dinner following, was to be held in Dallas, with her boss insisting she accompany him, as she had the year before. That time, she had not returned until well after ten p. m. There was no reason to think it would be different this year. The conference was scheduled for the first Friday evening in November, which would place it the week following Daily

Savings Time ending on the previous Sunday morning. Mickey would check with Jed and Rhett to see if that date would work for them.

Assuming he got the go-ahead, it would then be up to Jan, who was almost over her bug, with only sporadic coughs remaining, to do a sales job on their mother to allow the two of them to fend for themselves while she was in Big "D" for the night. On Thursday afternoon of the week preceding the week of his target date, Mickey and Jan spent a late afternoon in Heidi's home learning more of life beyond their planet before proposing his plan. It was agreed by all in the "family" that the chosen date sounded ideal.

"It should be nearing darkness by five-thirty p. m.," Jed pointed out. "If we can be in place at the transport site as near as possible to that time, we should be able to lift off, then chase the sun west, allowing a good view of the terrain in the light that remains."

Janice's eyes were sparkling with delight. "Where are you taking us? How far?"

"We do not wish to spoil all of your surprise. And there is always a chance we may deviate from the proposed path, based on weather and other factors."

Rhett spoke up. "Just be sure to arrive here by four forty-five to give us time to reach the site and prepare the transport. Do you think you can arrange that?" He looked at both siblings.

"Don't worry about that," Jan answered confidently for both of them. "We'll make sure of it."

"Fine. We will have the rental vehicle ready as before. By the way, we have decided that Jed and Heidi will be your guides. Jed, of course, is the pilot, and we thought you would like Heidi to share this adventure with you."

"Just the two of you?" Mickey asked, surprised, and a little disappointed for some strange reason. Glad that his friend would be joining them, he had assumed Rhett would be needed also.

"Yes," Jed replied. "The module only seats four comfortably. Blanche and Rhett will be monitoring our trip from here."

"Monitoring? How?"

"We will explain most of that before we depart that evening, and the remaining information will be better illustrated while in flight."

With goose-bumps of excitement and anticipation, the two young humans soon left the house, hoping the next week passed by in record time, which, of course, it would not.

Both sets of eyes kept a close watch for any creepy lawmen who might lie in wait as they walked.

Right on schedule, an hour disappeared from the clock on the following Sunday morning, and two a. m. became one a. m. The change had dusk falling an hour earlier than the day before. The earlier darkness made Halloween night seem more appropriately ghoulish for trick-or-treaters. Mickey, Benny, Trey, and Roby made a less than frightening quartet of Stormtroopers, the characters from *Star Wars.* Molded plastic and bent cardboard stood in for the real uniforms, all fashioned by Mrs. Payne, who had more time and resources at her disposal than the other mothers.

It felt good to have the gang together again, with the resentment existing between Mickey and Trey evaporating more every day. After-school meetings with Kristi had begun again, and that relationship seemed to be back on mostly solid ground, the only lingering scars being occasional questions from the girl about his frequent trips to Heidi's house. He did his best to laugh them off, assuring her that it was only friendship—nothing at all to be concerned about. The appealing girl's squinting eyes told him she wasn't completely buying it, but the promise of what lay ahead with Heidi and her fellow aliens wouldn't allow Kristi's doubts to cloudy his anticipation.

The chosen Friday finally came, with Mickey and Jan trying their hardest to conceal their excitement when sitting down to breakfast with Jessica. Neither sibling had any idea how to handle themselves as they considered the truth that by the end of the day they might be experiencing something that no one on Earth had before. It was almost too much to comprehend, and the fact an entire school day had to be completed first didn't help any.

"By the way," Jessica broke in to their thoughts, "I'll be coming home early to get dressed for the night. Should be here around four. Mr. Montgomery wants to leave the office by five or so. I'll head back to Mt. Pleasant around four-thirty."

Mickey and Jan exchanged quick glances. They hadn't considered that possibility. Their schedule may have just gotten a little tighter, time wise.

"Are you sure you don't want me to call Megan to come stay with you until I get home tonight?" Jessica asked, looking at her daughter. "I'd sure feel better about things." Megan was the infrequent go-to sitter.

"Mom," Jan answered with a hurt look on her face. "We'll be fine. I've got Megan's number. If things get spooky, I'll give her a call. But that won't happen."

"Well, I may call to check on you two tonight anyway."

Two nervous stomachs tightened up in unison. "Oh, Mom. Don't sweat it. Just have a good time, enjoy yourself. You deserve it." Jan showed a confident look.

That seemed to settle the matter, and the day that held wonder in store for two wide-eyed young Texans finally got underway.

At three-thirty, when the final bell rang, Mickey took a minute to explain to Kristi he had to do something with his sister that afternoon (which was the truth), then bolted to the line of buses to assure he got a seat near the front. He wanted to be the first off at his stop, hoping to have time to go over the plans with Jan for going to Heidi's house before their mother arrived.

The two conspirators had barely finished discussing things when Jessica Foreman pulled into the carport.

As Mickey and Jan had discussed, while Jessica was getting dressed in the only dress she owned suitable for the occasion, the boy left for Heidi's house to fill her crew in on the slight complication. His departure would be a normal thing, causing no suspicion in Jessica's eyes. Jan would join them as soon as their mother hit the road back to Mt. Pleasant.

The time at Heidi's before she arrived was put to use loading the rental car, this time a larger Ford sedan, with a few pieces of equipment Jed and Rhett had hidden under blankets. They were obviously some of the things Mickey had wondered about, hidden in the rear rooms of the house.

"Energy source, booster fuel, and other required parts," Jed said, nodding at the covered devices.

"Aren't they sorta small?" Mickey asked, eyeballing the cloth coverings.

"For human exploits, probably so," Rhett answered. "Remember, you have to adjust your thinking in regard to our technology, which has had centuries to develop beyond Earth's capabilities. If you recall, when we began our talks with you, we explained that our rate of speed is vastly superior to what you are familiar with on your planet. The same is true for all other scientific areas."

All Mickey could do was nod, having no reason to question what Rhett was saying. In addition to his excitement, there was no escaping a new tingling of fear—fear of what he would soon be part of.

Back at the Foreman home, Jan was growing antsy as four-thirty came and went. Her mother appeared to be in no hurry to leave for Mt. Pleasant, deciding between three pairs of shoes that she hoped would match her black dress. Nighttime events such as the looming meeting and dinner were few and far between for her, and she wanted to make this one perfect.

Finally at four-forty-five, she hugged her daughter and climbed back into the Toyota. As soon as she had cleared the intersection of Sabine and Tyler Streets, Jan threw on a jacket, grabbed Mickey's wind breaker since he'd left with nothing on but his shirt, and was on the run to her appointment with adventure, her lanky legs pumping rapidly. Noticing the chill-down in the weather conditions, she thanked herself for the jackets; temperatures were dropping rapidly.

When she arrived at 710 Louisiana Street, Mickey, Heidi, and Jed were sitting in the rental car, ready to hit the road as soon as she showed up. As she climbed into the rear seat beside her brother, Blanche exited the house, leaning in to address the siblings.

"Enjoy your adventure, Mickey and Janice. The weather is clear and cold, which should be ideal for the sights Jed has in mind for you."

As Mickey and Jan grinned excitedly at each other, Jed reminded them, "Blindfolds in place please. I apologize for this, but down on the floorboard as before, Mickey."

Jan tied Mickey's hood in place, with Heidi reaching back over the front seat to do the same for Jan. They then hit the floor.

"Well then," Jed once more announced, "we are off."

CHAPTER THIRTY-FIVE

After only a few minutes, it was obvious to Mickey that Jed was alternating the path of the car in the downtown area, circling the square clockwise instead of counter-clockwise as before. His body was shifting in the opposite direction on this trip through downtown; an apparent attempt to further confuse the route to the location of the transport. It was again accomplishing its purpose, for when the car made the final turn on the rural road leading back to the site, he was as lost as the first time. He couldn't hold in a hurt feeling, knowing they still didn't completely trust him. But again, he remembered what Heidi had said that first time, when Rhett had asked him to don the mask. The secret of the location of their transport vehicle had to be kept from him, and now Jan, for their protection as well as the alien visitors'. He had told Jan about all this, but he wondered how she was dealing with it, given her inquisitive nature. Their two bodies were thrown together from time to time as the side road became bumpy.

The vehicle again made the temporary stop to give Blanche and Rhett a chance to check out the security of the surroundings from the room in the rear of the house. Then it was on to the final destination.

"You can now sit up and remove the hoods," Heidi announced as the

Ford came to a final stop a few minutes later.

Mickey and Jan climbed out of the back seat, shaking the cramps from their sore legs.

Scanning the tree line circling the opening she now found herself in, Jan asked, “Where is it? I don’t see anything.”

“What did I tell you?” Mickey grinned. “You’ll see in a minute.”

“We must make it a quick minute,” Jed said. “We are slightly behind schedule, but should be able to gain some time if we hurry.”

With that, he went to the rear of the car, lifted the trunk and reached in for the items he had placed inside at the house. Leaving the blankets in the trunk, he reached in and handed one item to Heidi, who started toward the same section of the trees as the time before. Jed then followed with the two other pieces of equipment. All three were about the size of a small television set, steel gray, with what appeared to be terminals protruding from the shiny surfaces. Stopping roughly ten feet from where the pod had appeared as if by magic on the previous trip, he put the items on the ground, pulled out the same device as before from a pocket, and pointed it at the trees.

As before, the brush replaced the previous time fell away and the upper portion of the transport sprang to life. When it had reached the appropriate height, he then pushed a second button, and the bottom portion of the transport began to shake loose from its mooring. It continued to rise until it came to rest a foot or so from the upper part.

Jed approached, reaching in to once again disengage the charged electrical field. He then pushed another button that Mickey and Jan could not see, and a rectangular opening suddenly appeared on the top rounded shape. Motioning to Heidi, he retreated to her side and they picked up and carried the metal devices toward the pod. Three steps leading up to the level of the top section materialized as they reached the pod, allowing them entrance through the opening, which could now be seen as a doorway.

"Give me just a few minutes to install our equipment, then you can come aboard," Jed said, addressing the two humans standing and staring with mouths hanging open. Stunned into silence, they made no reply.

While Jed worked inside the pod, Heidi stepped out of the pod and returned to the car, driving it farther in to the trees, evidently concealing it until they returned from wherever they were going.

When she returned to join Mickey and his sister, she moved slowly, obviously laboring, wrapping her arms around her as she walked.

"You look cold, Heidi," Jan said. "Are you okay?"

"It has indeed become very cold," Heidi replied, her skin, or outer layer, looked as if it had tightened. "As we explained, the cold weather does not agree with our makeup. It restricts our movement somewhat. I will be fine once inside the transport." She handed the siblings the blindfolds she had retrieved from the car. "You may need these later. We will explain."

Mickey recalled that fateful day, now over a month back, a few days before he had seen something he wasn't meant to. Walking home that first afternoon, when spotting Heidi at her door, she had remarked on the chill of the Texas conditions.

Only minutes later, Jed appeared at the entrance way, beckoning to the three of them. "We are ready whenever you wish to join me."

Mickey was surprised when he found himself advancing slowly toward the steps leading up into the opening that had appeared from seemingly nowhere, feeling once more a sense of trepidation. Jan followed closely behind him; a clutch on the rear of his jacket with a trembling hand told him she was feeling it too.

Entering behind them, Heidi pressed yet another button, and the entrance way disappeared into the curved skin of the upper portion as quickly and silently as it had appeared. They all stood and stared at each other for an awkward few seconds, until Jed said, "Mickey and Jan, you have nothing to fear. In fact, we are sure this will be an experience you will long remember. A pleasant one, I am certain."

That seemed to have a calming affect on the two passengers. They peered around the interior, noticing two seats at one end of the module, in front of an instrument panel of sorts. Two more seats had risen from the bottom part of the craft, which were obviously meant for them. These seats were in the center of the floor, which still lay several feet beneath the upper section.

"Please, sit," Jed said, gesturing to the pop-up seats. "We will get underway now."

As they did, seat belts arose from below, wrapping around their mid-sections. They clicked into place snugly. Jan glanced at her brother, who attempted a brave smile in return, despite another tightening of his stomach.

Jed swiveled in his chair, facing them. He still wore his floppy hat, looking even more ridiculous than in the house."We will first rise vertically until reaching an altitude that will allow us to travel without causing alarm for any aircraft in the areas we pass through, probably just under 1000 feet. You will feel as if you are in an elevator, with only a small pull on your bodies. When we reach that height, I will put our transport in a horizontal speed which may cause motion sickness for you, since you have never traveled at speeds we are capable of.

"I will keep the observation windows closed, which will prevent most of your discomfort. When we open those windows to give you the views we wish for you to see, I will slow to a speed that will also prevent the loss of equilibrium, and will allow you a better view. In any event, if you choose to attach your blinders, you should be protected from the motion problems. Any questions?"

The two young humans were still too stunned to speak, shaking their heads in response.

"Well then, are you ready for us to begin?"

Excitement oozing from every pore of his body, Mickey blurted out, "I was born ready."

Jan turned to look at him, a sarcastic frown showing. Jed and Heidi looked back at the boy with puzzled expressions due to his comment, which was odd to them.

Jed then turned back and began manipulating a few controls on the panel facing him, waving a hand over the flat surface, then touching something. A faint sound similar to a vacuum cleaner on low speed could be heard, as the bottom part of the pod rose smoothly and silently as it joined the upper. A loud clamping sound indicated the two portions were coupled into place. A pleasant whoosh of warm air filled the interior of the vehicle, providing the perfect amount of heat. It felt good to Jan, and she could see Heidi in the chair next to Jed relaxing her thin body. Instantly, the sensation of rising from the Earth could be felt; sounds of limbs brushing against the side of the pod could barely be heard as the craft jettisoned straight up through the trees.

Mickey had been in an elevator only a few times in his life; twice while accompanying his mother, along with Jan, to an attorney's office in downtown Texarkana. It had had something to do with benefits from his father's death. He was very young then, and he could hardly recall that experience, but he was certain it was nothing like what he felt now. There were also a scant few visits to his paternal grandmother's fourth-floor condo in Rockwall, also very different from this. His body was pressed down firmly in his seat, but there was no pain or discomfort caused. The vertical rise didn't last long, since the speeds Jed had mentioned were true in all directions. In less than a minute, the momentum of the craft shifted to horizontal, and the stress on his body was gone.

If the transport vehicle was flying through the skies at speeds Jed had mentioned, there was no way to know. Having no need for it, Mickey shoved his blindfold into a coat pocket, then took a moment to peer around the interior of the pod, or rather, module. It was remarkably free of the many dials, metal tubes and other apparatus that was depicted on space ships and flying saucers in movies he had seen. In fact, the curved surfaces, except for the control panel Jed was monitoring, were remarkably bare. There were panels lining the walls, but if any controls were hidden behind them, it wasn't apparent. The immaculate appearance was creating a whole new round of questions he would pose to Heidi's family of aliens, but not until this promising adventure was done.

His mental list of questions was interrupted by Jed only minutes later.

"I am decelerating gradually since we are nearing the Dallas-Ft. Worth area. When we have reached a rate of speed that will make it possible for you to look downward without creating motion imbalance, I will open the observation windows."

"Wait," said Jan. "We can't possibly be near Dallas. When Mom drove us there last year, it took nearly an hour and a half. We've been in the air, what, less than ten minutes!"

"Remember what Jed told us about how much faster they can travel than us," Mickey pointed out.'

"Yeah, but . . ." She quickly seemed to just accept it as fact.

"Mickey is correct," Heidi added. "It is impossible for us to describe how our technology has given us such a tremendous advantage in relative speed."

Jed broke in to the conversation. "Here we are. I am staying south of the two cities, but you should be able to get a good view of the downtown areas of both."

Rubbing a hand over a dial on the panel, Jed activated controls which lowered opposite sections of the upper half of the pod, allowing a clear view from either side through curved, slightly tinted glass with no apparent distortion. At the same instant, all interior lights were dimmed, leaving just the panel lights on.

Jed explained, "We want you to see out, not for others to see in."

Releasing the seat belts, they both were able to move to the side facing the cities, observing in the sparkling clear night as dusk began to settle over the sprawling Metroplex, made up of the two large cities, plus hundreds of surrounding, smaller towns. The effect was spectacular.

"Wow!" the siblings said in unison. "That is amazing," Jan added. "Wonder which part Mom is going to be in tonight?" She thought for a second, then added, "Gosh, it'll be another hour at least before she and Mr. Montgomery even get there."

"More than that," Mickey said. "I don't think they've had time to leave Mt. Pleasant yet. Sure makes Cypress Springs look tiny," he added, trying to take the entire scene in.

Jed broke in to their reverie. "How would you two like to have what you might call a fly-over of DFW Airport?"

"That would be great," Mickey answered quickly. "Do we have time?"

"Yes. We will make a quick fly-by to the northwest. We cannot linger long there, due to the heavy traffic in the skies near the large facility. One short run perpendicular to the runways to avoid take offs and landings, then we'll be on our way to the next destination."

As the pilot made the necessary turn, Jan said, "You still haven't told us where we are going."

It was Heidi who answered. "Hadn't you rather it be a surprise? Sometimes anticipation can be the best part of a journey such as this."

"Well, I guess so." It was obvious Jan wasn't completely agreeing with that opinion, but let it drop. Mickey, however, agreed with the alien girl.

DFW was huge, separated into two main north-south sections of runways, with an expressway dividing it in half. The huge airport, situated between the two large north Texas cities, was only four years old, but tales of its enormous size had reached Cypress Springs. There appeared to be thousands of white, red, blue, green and yellow lights lining the runways, plus more lights emanating from the concourses and surrounding buildings. The gathering darkness added the right contrast to the spectacular light show below.

Just as he had said, Jed carefully monitored the commercial air traffic below as he cut across the center of the enormous complex. A new dial appeared on his control panel, which he explained was something similar to a radar screen, enabling him to judge the traffic below. Electronic voices came over a speaker located somewhere on the control panel. Jed's explanation for that was that he was able to tap in to air traffic control conversations with pilots in the area. In less than a minute, they were clear of DFW, heading west.

"Do you think we were spotted?" Mickey asked.

"Probably not," Jed said. "If we were, it will likely be assumed we are just part of the busy activity coming in or out."

That answer created another question. "Why are we not creating a fireball like the ones when y'all landed near Cypress Springs? I mean, we must not be making one, or else we would surely be spotted."

"Absolutely correct, Mickey. The reason for the trail you saw then was that we were encountering friction while re-entering Earth's atmosphere. We did not travel to Texas directly from Michigan. We had to rendezvous with others in orbit to explain our reasons for relocating. We then broke free of orbit, both nights, entering the atmosphere at tremendous speeds. If we had come straight from Michigan, we would not have been visible to you. For the same reason you usually cannot see commercial jets in the sky, which are at a much higher altitude than we are tonight."

Receiving nods of understanding from the siblings, Jed turned to the panel.

"I will close the window panels now, since we will be traveling at speeds much faster now. We have roughly a twenty-five minute flight. You should return to your seats and buckle up." As he started to close the panels the interior lights returned.

Mickey spoke up quickly. "Can you leave one panel open, so we can see the ground below?"

Jed hesitated a second. "I do not recommend that, but will do so. Be sure to keep your blindfold handy, just in case." The panel on Mickey's side was left partially open.

With a wave of a hand across the instrument panel, the transport accelerated immediately, the ground below becoming a blur of brown and green streaks, the hues becoming more vibrant as their path followed the sun, leaving the dusk of Texas behind. The blue-black air seemed to be enveloping the vehicle as it moved at speeds no human had experienced, without a hint of turbulence.

Mickey was able to tolerate the affect on his equilibrium for only a few seconds, then felt his stomach starting to heave. Looking away quickly from the window, he reached quickly for the blindfold, putting it in place.

Noticing his discomfort, Heidi reached across and waved her hand in front of Jed. The panel closed smoothly, and the foolhardy boy breathed deeply in relief.

CHAPTER THIRTY-SIX

Precisely on schedule according to Jed's calculations, less than thirty minutes later, the pod began to slow once again. He turned to face his passengers.

"Any ideas in regard to where Heidi and I are taking you? And, by the way, this first destination was Heidi's idea."

Jan looked at Mickey, who had removed his blindfold by now. Her face then noticeably brightened. "What state are we flying over now?"

"Arizona," Heidi answered. "Actually, approaching the northern part of the state."

"I know where we are going," Jan said excitedly. "Don't you know too, Mick?"

He shook his head. "How do you know?"

"U. S. Geography class. You studied this too."

Before she could finish, Heidi broke in. "Let it be a surprise for him, Jan."

The vehicle's speed was even slower now, and it was creating the sensation within the pod of gradually falling. When it finally leveled off, Jed opened the panels again, the lights dimmed, and sunlight streamed in filling the interior from the west, the overpowering glare making it hard to identify anything outside the windows.

"We are facing directly into the setting sun," Jed said. "I will adjust our direction so you can see better."

When he did so, the sun was at the pod's back, and the amazing scene outside became crystal clear.

"How about now, Mick?" Jan asked in a teasing manner. But her brother was too overcome to speak at first.

"Is this . . .is this The Grand Canyon?" he was finally able to blurt out.

"In all of its glory," Jed answered. "And I have to admit, there is nothing in our world that compares to it. We are just below the lid of the South Rim, on the southeastern edge of the canyon."

Mickey and Jan, now out of their seats again, leaned close enough to the window so that they could see downward as well as directly out. The sun was creating shadows in the bends of the huge rocks, eaten away over millions of years by the Colorado River and other forces of nature. A canvas of various shades of colors blanketed the walls; different colored layers of strata could be seen for miles. They were quiet at first, taking in a spectacular sight they never dreamed of seeing, at least until they were grown. The raw emotion of the experience had Jan putting her arms around her younger brother, glad they could see this wonder together.

"This is almost too beautiful to believe," she said softly, as if to herself.

"Is that a stream I see down there?" Mickey asked, pressing his head against the glass.

"More than a stream. That's the Colorado River, Mick," Jan said, showing she had paid attention in class. "It flows the entire length of the canyon, almost three hundred miles, as I recall."

"Actually, two hundred and seventy-seven miles, to be exact," Heidi said.

"Yes," said Jed. "And we are going to follow it the length of the Canyon, then on to our next destination. I will keep the transport in the reverse position so you can enjoy the view as we travel."

"How will you see where you're going?" Mickey asked.

"One of these screens on the instrument panel," he pointed, "shows a view from a camera located on the rear outside wall. There are others on both side walls. I can navigate in any direction I choose."

The boy simply shook his head. Nothing seemed impossible to these alien visitors.

As the transport began to trace the river below, the enthralled siblings clung to one window, then the other, taking in the width and depth of the yawning national treasure. The slow setting of the sun, now at their backs altered the shades painting the cliff edges, while steadily revising the shadow shapes. Various shades of purples and grays alternatively lengthened and faded. At one point, Mickey thought he could make out a small animal—perhaps a dog—climbing up a canyon wall.

"I bet it's a burro," Jan said.

"A what?"

"A burro, or donkey if you prefer," she said. "I've read they carry tourists up and down the canyon walls."

"You are correct," Jed said. "Here, let me give you a closer look." Running a finger over a screen, a camera mounted on the exterior was adjusted downward, then zoomed in for a clear view of the small pack animal, struggling with someone on his back, rocking awkwardly side to side, evidently trying to reach the rim before dark set in. Two more could be seen; one behind it and one leading the way.

"Unbelievable," Jan said, shaking her head again. "Mick, we've got to come back here some day and do that."

"Absolutely."

Traveling at a much reduced rate of speed to allow the two passengers to observe without becoming nauseous, nearly thirty minutes were required to reach the point where the Colorado exited the majestic canyon as it began losing its luster. The windows were then closed, as the speed increased.

"We are actually going to fly over Las Vegas, which is very near, considering our speed," Jed pointed out. "But first, you will be able to gaze down upon Hoover Dam, which has been built to use the Colorado for irrigation purposes. It is a remarkable achievement by you humans. In fact, we are approaching it now."

The speed reduced again, the windows re-opened. A new round of oohs and aahs followed, as Mickey and Jan could make out the enormous concrete structure they had both studied about. Behind the dam was Lake Mead, a sparkling blue reservoir formed by the Colorado River on the upstream side of the dam, as it snaked its way to the massive structure. Although it was obvious the host aliens were also impressed by the marvel of engineering and construction, their lack of emotions held their excitement at bay.

Hovering for only a few minutes over the dam, it was then on to Las Vegas, which took only five minutes, even at reduced speed. With the sun hanging low over mountains to the west, casting giant shadows over the spread-out town, there was still enough daylight to observe the sights of Sin City, as the desert town was referred to.

"I will parallel the street known as the Strip, but is actually Las Vegas Boulevard," Jed said. "I will descend as low as I can to give you better views. However, there is a chance we may have to leave the area on short notice."

"Why is that?" Mickey asked.

"There is a large Air Force Base nearby, Nellis Air force Base. Any unusual activity in the skies in this area may arouse their curiosity."

"Well, I think your pod, or module, would definitely fall into the unusual category," Jan said, stating the obvious. She seemed rather excited about the thrill of being discovered. "But how will you know if they're alerted?"

"I can access their frequencies and monitor any concerns they have. It is also entirely possible someone on the ground will spot us at our low altitude."

With that danger hanging out there, they applied their attention to the sights along the Strip, which was teeming with activity. The city proper could be seen in the distance, but all the action apparently was here, on the busy strip.

“Look,” exclaimed Jan. “There’s a giant cowboy in front of a building, waving his hand back and forth.”

“A casino,” Heidi pointed out. “Several of them use such items to draw customers in.”

“I see a giant silver slipper,” an excited Mickey said. “And a huge circus clown in front of *that* casino.”

Heidi nodded. “Those are the Silver Slipper and Circus-Circus casinos.”

Mickey started to ask how she could possibly know that, but then, these beings seemed to know everything. With the odd things he was viewing from above on the long boulevard, he thought it would be next to impossible to notice the craft they were in, seemingly floating over the people down below. It should blend right in.

Jan could make out the names of celebrities she recognized on a marquee. “Wow. Dean Martin and Sammy Davis are performing at that one.” She continued looking. “And Wayne Newton at that one.” She then remembered something about one of her favorite performers. “Didn’t Elvis perform at one of these places before he died?”

“Yeah, that’s right,” echoed Mickey.

“We recognize those names,” Heidi said. They were all entertainers, as we understand. Were they important to you?”

“They were to me,” Jan said. “The ones still alive still are. Why?”

“We don’t have that sort of . . . I guess you would say, celebrity, on our planet. I think that is unique to Earth.”

“Is that good or bad?”

"Neither. It is simply different, and the type of thing we wanted to discover about your world."

Mickey looked at Heidi quizzically. "Do you have any entertainment on your planet?"

"Yes. But ours is more in the form of dramatic presentations. Plays and stories our ancestors have written. Music, as far as we can determine, is something else that is distinct to your world. We rather enjoy it."

Jed nodded his agreement from his seat.

The sounds of Lawrence Welk's odd tune the "Jones" danced to weeks ago instantly came back to Mickey.

Jed started to add something else but suddenly sat up. "Hold on, we have company. Time to go."

"What? Why?"

"Fighter jets are being scrambled from Nellis. We have been spotted. Please take your seats and buckle in. I am closing the window panels. Our return to Cypress Springs will be non-stop at very high speeds. These fighters might alert other bases along our path, so we must hurry."

After executing a quick u-turn high above Las Vegas, while simultaneously rising to a higher altitude, Jed spoke something to Heidi in that strange garbled sound Mickey had heard a few times. It was plain to see the pilot was concerned. Even buckled in with the window panels closed, Mickey could feel the speed Jed referred to after the pod leveled off. He felt his body pushed back against his seat as the craft soon hit speeds that had to be enough to break the sound barrier several times over. In fact, the earlier speeds reached on the initial portion of the trip should have produced a sonic boom, based on what he had read. It was another question for the pilot, but it would have to wait, for he was now having to concentrate on leaving the US Air Force jets far behind.

CHAPTER THIRTY-SEVEN

After a brief passage of time, less than five minutes in total, Jed seemed to relax, evidently confident he had left his pursuers far behind. From time to time, the pilot spoke his electronic sounding jibberish into the panel, receiving return messages, apparently from Blanche or Rhett, sitting in that mysterious room on Louisiana Street back in Cypress Springs, which in itself was mind bendingly surreal. The strange voices came through as clearly as if they had come from within the pod.

Mickey took advantage of the relief of tension in the transport to pose his question about the lack of any "booms".

"You are right again, Mickey," Jed said, as he swiveled once more in his seat to address the boy. "We cruised through the sound barrier, which is reached at the same speed regardless of where you are—your world or ours—twice during our flight. Somewhere between the Dallas-Ft Worth area and the Grand Canyon, and again when we left Las Vegas minutes ago. Sonic booms, as you refer to them, occurred both times, but were only heard on the ground below."

Noticing a puzzled expression, he continued. "You see, at elevated rates of speed such as ours, almost the second we encounter the wall of pressure representing the sound barrier, we are through it and beyond by quite some distance, leaving the occurrence behind. The disruptive waves we generated by passing through and disrupting the barrier travel outward, including downward to Earth, creating a burst of energy represented by sound bouncing off the surface. It is likely citizens below at those points began talking to their neighbors, comparing notes, wondering if someone might be detonating explosives in the area."

Mickey looked at Jan, gauging her reaction to Jed's explanation. She simply shrugged, understanding even less than her space-struck brother. For his part, he thought it made sense, or as much as anything else would. All four passengers remained belted-in for the rest of the thirty-five minute trip back to northeast Texas. Although he could sense the enormous speed of the craft as it hurdled through the skies, there was amazingly little discomfort caused to Mickey's body. The sound of air whooshing over the outside layer of the pod was so faint as to barely be heard.

It was seven-forty-five by the time the craft touched down gingerly in its unknown resting place somewhere near Cypress Springs. Mickey had hoped the window panels would be opened as they landed, so he could get his bearings in relation to downtown. Jed, however, was still guarding the secret of the hiding spot, keeping the panels closed.

Things were then done quickly in reverse order from earlier; the lower section lowered in place, as it separated from the upper, before brush was replaced covering the upper shell, leaving it as they had found it hours before. Heidi then retrieved the rental car from its hiding place. Finally, with a brief garbled message to the watchers at home, evidently giving notice of a successful return, blindfolds were back in place, and the ride into town began. It didn't occur to Mickey until he thought they were nearing the city limits, that Jed and Heidi had failed to remove the pieces of equipment they had brought from the house for the ride home. He had no idea what that meant, but knowing their reasons for doing everything they did, he quickly let it go.

Twenty minutes later, when the Ford had reached their neighborhood, Jed offered to drop the two humans at their home. Fearing the possibility of Jessica returning earlier than expected, Jan said, "No thanks. We'd better walk from your house. It would be tricky explaining to Mom why we were in your car."

The walk home was a slow, quiet one, as the adrenalin of the past few hours began to ooze from two young, keyed-up bodies. The amazing excursion had lasted less than three hours, including the time spent in the car, but had shown them glimpses of wonder they could've hardly dreamed of only days ago. Now, the aftermath was hitting them hard. The recurring thought seemed to hit them both simultaneously; the unfairness that the adventure would remain known to no one else, only to the two of them, was a depressing reality. Not even their mother could know. Mickey broke the silence as they reached the corner of where their street intersected with Louisiana. He came to a stop.

"Why me, Jan?"

"Whatta you mean?" She gave a somewhat puzzled look, but Mickey was sure she understood what he was asking.

"Why did Heidi and her family, or group, or whatever, choose me to tell their secrets to? They coulda picked anybody. I mean, I'm glad they did, 'cause I've enjoyed every single minute of it, but I just don't understand."

Jan thought for only a second, then gave an answer she saw as the truth. "Because you are trustworthy, Mick. I think people, and even aliens I guess, can tell just by watching you that you won't betray a trust. And the way you warmed up to Heidi when no one else would showed them what you're made of." She paused, gazing down at her little brother, who seemed lost in thought. "You're just a good guy, Mick. At least deep down you are. On the surface, you can still be a twerp at times." Grinning, she gave his shoulder a bump with her own.

"Look, don't stress over it. I've enjoyed it too, and you have to accept it as their way of discovering friendship on our planet. That may not be why they are here, but it's what they found with you."

"Do you think we'll ever be able to tell anybody what we've seen and heard?"

Pursing her lips for a second, she responded, "I have no idea, but I'd like to think so. One thing I do know is you and I will be telling our grand kids about this, and we can always share it with each other."

The nod he gave in return was accompanied by a touch of sadness.

The phone was ringing as Jan unlocked the front door minutes later. She rushed to the den to answer it, knowing it could only be one person. Mickey watched with concern as she picked up.

"Jan? I was about to hang up. What took you so long to answer?" Jessica asked in a worried tone of voice.

"Oh, hi, Mom. Mick's napping, and I guess I had my music on too loud. Sorry." Trying to keep her voice at a normal level, Jan exhaled while putting a hand over the mouthpiece, then asked, "How's the meeting going?"

"Fine. You sure you're okay?"

"Yeah. Nothing happening here." She exchanged grins with her brother.

"Well, okay then. We're cutting the dinner short, be heading home soon. Should be there by ten-thirty or so."

"Okay, Mom. Be careful driving back from Mt. Pleasant."

Hanging up, Jan fell into the recliner. "That was close."

Later, as Mickey and Jan continued to wind down from the magical flight across half the U. S., the Foreman family sat in the living room, watching with detached interest the late news piped in from a Shreveport tv station. Jessica and her trim daughter were wedged together in the recliner, with the girl's head resting on her mother's shoulder. Mickey was stretched out on the sofa, doing his best to stay awake. He was almost afraid to nod off, and definitely apprehensive over retiring for the night. There was a chance, he thought, he might wake Saturday morning and find it had all been a dream. Earlier, before Jessica had arrived, he had pulled an Atlas from the book case, opening it to a fold-out map of the country, tracing with a finger the path taken by the transport. The distance to Las Vegas and back measured out to be just over 2200 miles. Including the time spent meandering along the Colorado River, and the fly over of the Vegas Strip, plus the brief detour over DFW, those miles had been covered in slightly over two hours. Impossible!

As the blow-dried news anchor got off the subjects of burglaries, assaults, and the cost of a gallon of gas, changing his tone to something lighter, the siblings came to life. Jan raised her head and Mickey sat up on the sofa as the next story began:

"All of you Star Wars fans, listen up. Possible UFO sightings are being reported from out west. A group of tourists are saying they saw something odd hovering over The Grand Canyon late this afternoon. Varying opinions state that an object resembling a giant silver ball, a shiny cigar, and even a utility van, could be seen. It was spotted by several other groups at more than one location in the canyon.

"Roughly one hour later, similar sightings were reported over Las Vegas. Officials at nearby Nellis Air Force Base apparently picked up strange activity in the area, putting jets in the air to investigate. They were unable to identify any such suspicious appearing sightings. In fact, they saw nothing, labeling the incident as possible weather balloons."

The anchor put on his best dramatic face, adding,

"I don't know about you, folks, but it seems odd to me two such sightings within such a short time, in two different locations, can be laughed off as weather balloons. What do you think?"

Jessica shook her tired head, yawning. "I'd like to tell him what I think," she said. "People are desperate for anything that seems to indicate there is life somewhere out there. I can't believe how gullible they are."

Launching a grin in her brother's direction, receiving a tired one in exchange, Jan climbed from the recliner, giving her mother a kiss on her cheek.

"Well, Mom. You never know." Another yawn and a stretch, and then, "I'm going to bed. It's been some kinda day."

Jessica didn't react to the implied bait. "Good night, honey."

"Me, too," Mickey said. Planting his own kiss, he followed his fellow explorer down the hallway.

What neither the news anchor, nor a certain young brother and sister in Cypress Springs, Texas, knew, was that an American Airlines pilot had also spotted something suspicious earlier in the afternoon streaking across the runways at DFW. After giving an initial report to the control tower, and being asked if he wanted to make an official report, he sighed, saying emphatically, "Negative. I do not." Following a long day in the air, he was in no mood to deal with what would ensue after filing such a report.

CHAPTER THIRTY-EIGHT

On late Tuesday afternoon of the following week, an awkward, rather brief meeting took place in the private office of Police Chief Winston Dunaway. The mood of the meeting had to do with the fact two of the three people involved weren't sure why they were in attendance, and had rather not have been. Their names were Lewis Medlin and Jonathan Crowley, both agents with the Immigration and Naturalization Service, an agency now under the arm of the Justice Department. The service was referred to as the INS; like most federal agencies, they preferred acronyms to save time.

Introductions were made by Police Officer Brad Sweat of the Cypress Springs Police Force, at least in the case of Crowley, the officer he had been in training with in Dallas. Crowley, in turn, introduced Medlin to Dunaway. It was Crowley who appeared to be the more senior of the two agents, although both were young, no more than their early thirties. Sweat excused himself after performing his duty, having been informed by his boss earlier he would not be needed during the actual conference.

The age of the two men left sitting across from him was one reason Dunaway was disappointed by what he saw. He had hoped they would've been more seasoned—meaning nearer his own age, which would indicate they had a better view of the real world—meaning they might be inclined to see things as he saw them. They both appeared to be in good shape, about six-two, with flat bellies befitting of younger law enforcement types. He was sure those bellies would begin to extend outward after a few decades of chicken-fried steak and Tex-Mex, if they remained in Texas.

The other thing that bothered him was that one of the agents, Medlin, was black. If Dunaway were ever asked if he was prejudiced toward people of color, he would scoff and answer, "not a bone in my body." If citizens of his community that knew him at all were asked that same question about him, their replies might be, "there could be something to that", followed quickly by, "but don't quote me on that".

But they were here now, having driven five hours from their office in Austin, so the police chief had no choice but to get whatever he could from them.

"What can we do for you, Chief Dunaway?" Crowley broke the ice. "Brad was a little vague on the phone call last Friday. Fill us in."

Dunaway leaned back in his chair and folded his hands in his lap. "That's because I didn't tell Officer Sweat about my concerns. Didn't want him to be spreadin' rumors about the people I would like you to take a look at. At least, not until you've had a chance to form your own opinion."

"Makes sense. Go on."

The chief spent a few minutes explaining the efforts he had made to uncover the story of the Jones family of 710 Louisiana Street, concentrating on the fact no social security numbers had been found, and no record of any kind as to the family's actual existence, including no phone listing, the one exception being the lease agreement. He threw in for emphasis the fact they had no friends or relatives that brought them to

Cypress Springs, nor a job awaiting them here. The vagueness of their previous residences was only due to the fact no one actually asked them more about that, but Dunaway added it, giving the impression the family had tried to hide that history.

The agents stared across when he was done, blank expressions saying they didn't get it. They shared doubtful glances with each other, then Medlin finally spoke. "Could you explain how they came onto your radar screen, and what made you begin your search?"

This was the tricky part. Dunaway shifted noticeably in his chair. "I've had three citizens," he held up three digits for effect, "ask me about them. Said they were neighbors of the Jones', but were becoming suspicious about them. Of course, I tried to convince them there was nothing to worry about, but they seemed genuinely bothered by how secretive that family is.

"Now, they got a daughter that goes to school, who they say seems fine. But the adults are another matter."

"Do you have any way of knowing of any speech dialects they might have? Hispanic, or anything else?" Crowley asked.

"Haven't had the occasion to hear them speak. Don't think the neighbors have either, except for the girl. She doesn't seem to have any noticeable accent, at least, according to those folks."

The agents nodded their heads slowly, but otherwise, nothing.

Getting no discernible reaction from the men, Dunaway decided to play his ace, hoping to create some form of interest. He was becoming desperate, knowing he needed justification for requesting the meeting. In fact, he was starting to wish he'd never brought these two guys here.

"Crying wolf" to a federal agency was no way to enhance his reputation. So, he chose a lie that should generate some sort of reaction.

"The other thing is, and this is what really concerns those neighbors, and me too is, there are some kids in their neighborhood who have been spending a lot of time with these individuals in their home. I mean, why would those folks, who don't know a soul in town, and with no apparent reason for moving here, have those kiddos in their house? And they are good kids—I know the family well—the mother is a good friend of mine." The deception in that last statement caused him to again shift his large girth in his chair.

Crowley pursed his lips, nodding slowly. "That does come off as strange. Have you asked the mother why they are in that house so often?"

The question caught Dunaway off guard. His story didn't contain an answer for that.

"Well, I simply didn't want to alarm her. Being a widow, she's swamped enough with raising two kids alone." He grinned a toothy smile while leveling his floating irises at the agents. "I'm sure you can understand that."

Whether they did or not, neither man responded right away. They again shared a glance. It lasted for a few seconds, each apparently using some sort of telepathy to gauge their partner's thoughts. It was Crowley who finally said, "Well, Chief, it all sounds a little thin to us, but we do hear you. Give us a minute or two to compare notes about this."

He rose from his seat as Medlin followed suit. They then left his office, then the building, without saying another word. When they had closed the door behind them, Dunaway waited a moment, then followed.

He casually strolled into the modest reception area, peering through the large window facing the front parking lot. The agents were standing at the edge of the lot. Crowley's booted foot rested on the bumper of a black Ford sedan with a INS emblem on the front doors. They appeared to be arguing the point. From his vantage point, it looked to Dunaway that Crowley was favoring humoring him, while Medlin continued to shake his large black head.

Finally, the decision was apparently made, and the agents headed for the front door of the building. Dunaway hurried back to his office as they approached, preparing for whatever they might say. As he waited, he knew well if he had described the encounter he had experienced with one of the people in question, it would get immediate reactions from these federal guys. But he couldn't mention that without explaining how the incident came about.

"Okay," Crowley said as he closed the chief's office door behind him and Medlin. "We'll pay a visit to these folks, feel the situation out. If we are left with the same suspicions your citizens have, we'll investigate further. If not, which is more likely the case, we'll sign off. Hope you can understand that."

"Certainly."

"Okay then. Give us directions to their house and we'll make a drop-by."

Scratching out the address, along with a map to their home, and the family's name, or at least the one they were using, he handed it over to Crowley, saying, "Sure appreciate you guys coming all this way."

"Right," Crowley replied, reaching across the Chief's desk to shake. Medlin followed suit. "We'll be in touch," Crowley added.

CHAPTER THIRTY-NINE

Mickey sat and watched as his sister and his alien classmate came as close to bonding as they ever would, or could. Heidi, of course, wasn't capable of developing feelings of any sort toward humans, or likely any other species. But over the course of the sessions spent in the Jones' household, and especially during the unforgettable flight the week before, he had seen Jan warming up to the "girl" she had labeled a freak only weeks ago. They had shared what music meant to them both, but obviously from different perspectives. When asked why the Lawrence Welk music had been chosen for the unearthly "dance", the surprising answer was that it matched the attitudes of natives on their planet better than other music on Earth. That reply generated a discussion between the music lover on Planet Earth and the being from a different world that caused Mickey to zone out while they talked.

Among the other sad realities of the secretive relationship he and Jan possessed with the "family" residing at 710 Louisiana Street was that it was not a real friendship, and never could be. Yet, he knew without any doubt, they in some ways, were the best friends he had ever had in his still young life, even though he had known them for just over one short month. They never made judgments, simply because they weren't capable of doing so. All they did was share what they had—which was knowledge beyond anything he had thought possible. The hard fact that it was knowledge that only he, and now Jan, carried with them, kept him awake many nights.

The three of them sat at the modest dining room table on late Tuesday afternoon, with Jan holding the puzzling phone device in her hands. She had never had the chance to put it to use, and she was getting a kick out of calling a couple of her friends. Speaking their numbers into the mouthpiece without dialing was something a budding teenage girl saw as miraculous. Her friends on the receiving end had the same reaction she had that first night when Mickey had called. They both said it sounded as if she was standing right next to them. Heidi smiled faintly at the thrill the communication device that was commonplace to her was providing Jan. The alien girl, none-the-less, closely monitored the conversations, not wanting information about the location or source of the calls to slip out.

The night before, Monday night, Mickey had managed to force the courage up to ask Jed and Rhett if an encore flight might be a possibility, fearing the answer he might receive. He still couldn't let go of the fact the equipment needed to operate the craft had been left on board when returning from the first flight. He took it as a hopeful sign they may be considering returning for a new adventure. *Why else would they leave it there?*

"We will see," had been Rhett's reply. "It is a possibility."

"Oh. Okay."

Evidently recognizing disappointment, Rhett had added, "It depends on several factors, none of which we can predict or control."

That confusing answer had left things in doubt, but Mickey had let the subject drop, knowing it would be pointless, and likely rude, to press any further.

Perhaps as a consolation of sorts, Jed had lead him back to the mystery room in the interior of the house. "We have told you about many of our devices," the floppy-hat wearing alien had said. "There is no reason to keep more secrets."

Mickey followed anxiously, having no idea what to expect to see. Walking past two rooms in a short hallway with their doors closed, he assumed they were bedrooms. *Even that* lead to other curiosities: did they all sleep in those rooms? And did they sleep in beds like humans, or hover, similar to what they did when they performed for Jan and him? Heidi and Blanche in one room, and Jed and Rhett in the other? He did his best to ignore those concerns, aware there were some questions he might never receive an answer to.

The so-called equipment room proved to be something of a disappointment. There were three screens appearing to be large tv screens, which Jed referred to as monitors, sitting on a table next to the far wall. Viewing them, Mickey could see the scenes of the hiding place somewhere outside of town, plus the roadway leading back to it, as recorded by the cameras positioned in trees. A speaker, and what appeared to be a microphone of sorts, though nothing like the ones in school assembly, was anchored to a small desk. Other than that, just a few metal bins where the equipment left on the transport vehicle had been kept. Apparently noticing Mickey's underwhelmed reaction to what he was seeing, Jed pointed out, "Keep in mind, Mickey, with our technology we can do a lot with what appears to be a little. It has taken many centuries to perfect our devices, including adapting them for use in far away locations. I can see you are somewhat disappointed."

"No, sir . . . I mean, Jed. It's just that, well, I wasn't sure what to expect to find back here." He walked closer to look at the images on the screens. "So, Blanche and Rhett could see us while we were at the transport site?" He chose to use the term they seemed to prefer, to make up for his show of being disappointed.

"Exactly. However, we mainly used this arrangement to observe the security of the location we chose to conceal our transport module. We are sorry to have been so secretive in not providing access to this information to you before. We simply must be extremely careful in not exposing our existence and procedures. I hope you can understand."

Mickey nodded quickly. "Yes, I do. Completely. Thanks for letting me see it now." That had ended the impromptu tour.

Still thinking about what Jed had shown him and explained the night before, Mickey turned his attention back to Jan and Heidi bonding over music.

Suddenly the three other members entered the living room hurriedly from behind the wall. Blanche and Rhett took stances in the middle of the room, staring hard at the front door, while Jed stood next to Heidi, who had stood rapidly from her chair. They also were staring at the door.

Mickey looked toward Janice with a questioning stare. She shrugged her shoulders in response. Something was bothering the residents of the house, something the siblings could not detect.

"Mickey, Janice," Blanche said abruptly, "it will be better if you two move into the kitchen."

"Why?" Janice asked. "What's going on?"

It was Rhett who answered. "We sense someone approaching, and not knowing their purpose, it will be simpler for us to discover it ourselves. It will be safer for you to keep your presence hidden until we determine the situation."

With the siblings showing hesitation, Heidi addressed Mickey. "Please, Mickey. Trust us once more."

The two did as requested, entering the kitchen, seating themselves in vinyl-covered dinette chairs next to a small round table placed against the wall separating the kitchen from the larger room. Leaning in to the wall, they trained their ears on the sounds from the adjacent room. A loud knock could be heard at the front door. It was Rhett who answered the knock.

"Good evening. May I help you?"

There was a slight hesitation before the visitor responded. "Mr. Jones? Rhett Jones?"

"Yes."

"Well, okay, sir." A throat could be heard clearing. "Mr. Jones, my name is Jonathan Crowley, actually, Agent Jonathan Crowley. This is my partner, Agent Lewis Medlin."

A new voice could be heard, saying, "Good evening, Sir."

"And to you also," Rhett replied.

A brief pause, then, "Agent Medlin and I are with the Immigration and Naturalization Service, better known as the INS," the first voice spoke again, then another gap in the conversation while badges were produced and displayed.

Behind the wall that separated the rooms, Mickey looked at his sister, mouthing “What is this?”. She answered with a shake of her head, putting a finger to her lips.

“Oh? I see.” Rhett’s voice seemed calm, yet tinted with a degree of doubt and confusion. “Is there something we can do for you?”

“Well, we were hoping so. Do you mind if we come in for a second? Shouldn’t take too long. Just a few questions.”

“Why, of course. Please come in.” Rhett could be heard introducing Blanche, Jed and Heidi to the agents, identifying Jed as his father. There was a shuffling of chairs as two dining room chairs were pulled across the room.

Crowley and Medlin were offered the sofa, while Rhett and Jed took the wingbacks. Heidi and Blanche were seated in the dining room chairs. An awkward pause followed, then comments about the weather and how nice the town of Cypress Springs appeared took up a few minutes of warm-up time.

“Er, we really don’t need your daughter to participate,” Crowley said. “Wouldn’t want to upset her.”

“Oh, Agent Crowley, we have nothing to hide from Heidi,” Rhett said. “There are no secrets here.”

Crowley looked at the girl, offering her a smile he intended to be non-threatening. “Okay, then. That’s fine with us.” He glanced sideways at Medlin, who raised his eyebrows and nodded.

“To begin,” Crowley said, “we apologize for showing up unannounced like this. You look like nice folks to us, so please keep in mind we don't mean to pose as threats to your family, but we do have our duties to perform. It's just that, well, some local citizens have expressed concerns about your, uh, presence in their community, and they have asked us to make inquiries."

"Concerns?" Blanche interjected. "May we ask what kind of concerns?" She leaned forward, which she saw as an appropriate reaction to the man's statement.

Crowley waved a hand in her direction, as if to allay her fears. "We'll get to that in a minute, ma'am. But first," he said, looking again at Rhett, "would you mind telling us where you came from when relocating to Texas?"

Rhett shared a curious look with Blanche before answering. "Why, we moved here from Michigan . . . the Upper Peninsula, to be exact."

Crowley nodded slowly, as if he had heard of that part of the Wolverine state, but wasn't quite sure about what it consisted of. "And your reasons for choosing to reside in Cypress Springs? I mean, in particular?"

Rhett managed to form a frown across his brow, which took some effort on his part. He shrugged. "We just saw it as a pleasant place to live. Not too large, not too small. Fairly close to the Dallas area, as well as Texarkana and Shreveport."

Jed finally joined in. "I can't imagine why anyone in this community would be at all concerned about where we came from, or why we decided to make this our home."

Medlin now leaned in from the sofa, resting his elbows on his knees as he spoke. "You know, sir. I can't either. Sometimes folks can read too much into the life of strangers, without knowing anything about them." As he voiced the words, he couldn't remove his eyes from the silly floppy hat the man wore, actually smiling at the sight.

Crowley shot a look at his partner. "Never-the-less, we owe it to those other folks to gather a little information, if for no other reason than satisfying their curiosities. We, for example, were wondering if you would mind sharing your I. D. numbers with us?"

"Our I. D. numbers?" Rhett now tried out his confused expression.

"Yes, sir. Your Social Security numbers."

Rhett sat back in his chair, spreading his arms apart. "But Agent Crowley, we do not have those numbers."

Crowley stared at the man for a second. "You don't have . . . everybody has to have a Social security number." The agent was getting a little agitated. Medlin also had a frown developing.

"But we do not have them, because we are Canadian citizens. We received assurances from both our government and yours that, as long as we didn't hold jobs in the U. S., it was not necessary to register with the Social Security Administration. Therefore, we have not done so."

Crowley's mouth hung open for a second. "Canada?"

"Yes, that is true. I am sorry if I have confused the issue. You see, we first moved to Michigan several years ago. One reason we did so was to escape the brutal winters of the far northern part of Quebec Province. Unfortunately, Upper Michigan was also cold, so we relocated farther south."

"I think," Medlin said, glancing at his partner, "that what Mr. Jones has said is correct. The U. S. does not require Canadians to obtain a Social if they are not employed."

Crowley let a small amount of air escape from between his lips. "If none of you have jobs, how to you handle your living expenses?"

In the other room, Jan smiled, while tapping Mickey lightly on his shoulder, recognizing one of her own questions posed during an earlier meeting.

"We have had jobs, all three of us . . . my wife and my father and I . . . saved during our working years. That has enabled us to retire and allows us to maintain a life style we feel comfortable with."

Crowley seemed frustrated, his head down as he searched for another question. He suddenly looked up and asked, "What was the city, or town, in Quebec you were native to?'

"Why, Gatineau, in the northern section, as I mentioned."

"I see. Tell me a little about that town, if you don't mind." The question had the clear tone of an accusation attached to it.

"Gatineau? Well, it's a fairly large city, with a population of somewhere around 200,000. It sits on the banks of the Ottawa River. It is a beautiful area, but as I mentioned, rather hard to endure in winter months."

Crowley was plainly losing whatever personal battle he was having with the man who was remaining remarkably calm. He appeared ready to test the man's story with more questions, when Medlin spoke up.

"It does sound like a nice place, but I can't argue with moving to a warmer climate." The black agent was aware his partner was staring daggers at him from the other end of the sofa, but he felt they had gone far enough and he wasn't about to let things go too far, if the only justification was to satisfy Chief Dunaway, who, regardless of what he had said, he was sure had his own motives in regard to these non-threatening newcomers.

"Partner, I think we can let these folks be. It appears to me . . ."

Medlin's statement died in mid-sentence, as all four of the people sitting across from the agents suddenly rose from their seats.

"Are there others outside?" Jed asked, addressing the agents.

"Others? What others?" Crowley said brusquely, still smarting from his partner's remarks.

"Agents, like yourselves. Are there more agents outside?"

Crowley and Medlin looked at each other. "No," Medlin answered. "We came alone. Why?"

Before anyone could speak, voices from outside could plainly be heard; loud voices, yelling at each other, although the words weren't intelligible. The agents also rose to their feet. When the sound of a gunshot pierced the momentary silence, the agents instinctively reached inside their jackets.

"I swear," Medlin said to the family, "this is not our doing."

The agents approached the door.

CHAPTER FORTY

Chief Winston Dunaway was just suspicious and paranoid enough to not trust anyone who might jeopardize his career. Especially not two out-of-town INS agents who likely cared little about his future one way or the other. Allowing them a head start of roughly a half-hour, he drove to a spot two doors down from the suspect house and parked. Spotting the Ford with the federal agency insignia on its doors at least told him they had begun what he hoped was a thorough interrogation. He decided to wait it out, wanting to de-brief them as they left the interview.

After fifteen minutes, his legs and butt started to cramp up. Climbing from his vehicle, he stretched in the gathering dusk. It was a clear, cool early evening, and he felt good; maybe things were turning in his direction. He paced a few yards back and forth along the sidewalk, trying to stay patient. The fact the discussion inside 710 Louisiana Street was taking so long was another good sign. They were no doubt uncovering dirt on the mystery family; enough dirt to give him reason to exact his revenge.

His pleasant thoughts were interrupted by the loud sound of a door being slammed shut, somewhere in the film of early evening. His back was to the location of the noise, giving him no clue as to which front door it emanated from. He had a sickening feeling he might know the answer to that question, and that feeling quickly became a tight knot in the pit of his stomach. The outline of a large figure came rumbling toward him, swaying noticeably side-to-side on the sidewalk. Dunaway's fear of all fears came into focus as Jim Robinson marched unsteadily in his direction.

"Dunaway, you bastard! You and my wife?!"

A shock of pure panic enveloped the lawman, causing him to forget completely about the meeting going on in the house in front of him. He managed to find his voice.

"You're drunk, Robinson. You don't know what you're talkin' about."

"The hell I don't. How in hell did you think you have the right?!"

"You're crazy, Jimbo. You know me better than that." Then, as the man continued marching at him, he put his hand on his holstered hand gun as he tried to think of a way out. "What did she tell you? What kind of lies?"

Robinson was now only fifteen feet away. Spittle flew from his mouth as he spoke. "It wasn't her, you asshole. It was my son! My son! He told me! I've taken care of her. Now I'm dealin' with you."

Dunaway drew his handgun, backing up a few steps. Wondering what the beefy man had done to his wife, there was no telling what he had in mind for *him*. He pointed it at the raging man. "Don't come any closer Jim. I'll fire. It'll be clear self defense."

"You coward! You're going to eat that gun. It can't protect scum like you."

The man kept coming, drawing within ten feet. Dunaway fired, hitting the charging man in the right shoulder. The discharge echoed through the night in the neighborhood. Robinson jerked his body to the right, absorbing the round, but quickly resumed his approach. The lawman stepped back another few feet, alarmed at the man's ability to withstand the shot. He fired again, just as the front door of 710 Louisiana Street flew open, and two dark clad men with weapons drawn came rushing out.

When Crowley and Medlin exited the Jones' house with weapons drawn, it took a minute to take their measure of what was happening. Inside the house, they hadn't been able to grasp the words exchanged in the confrontation between the two men, so their identities were still unknown. In the dimming conditions, all they could make out was a man holding a weapon, a weapon he had fired at an individual who was now lying prone only feet away from him. Smoke from the gun was easy to detect in the filtered light.

"Drop your weapon!" Crowley yelled. "Down on the ground! Now!"

The man was still facing his victim, with his right side exposed to the two men pointing their guns at him. Confused as to why they would be drawing on him, he had trouble speaking, or thinking. Why were they barking commands at him? He was the Chief of Police of this town—his town. No one questioned his right to defend himself. Unable to realize they had no idea who the man was that was holding a gun he had just fired at someone lying in front of him, he slowly turned toward them, forgetting his gun was in his right hand. It turned with him, which was all the threat the two agents required.

Opening fire on the man, Crowley and Medlin recognized too late the khaki uniform of Winston Dunaway. Medlin's round landed in the chief's right shoulder, with his partner's hitting just below the right hip. They could actually hear bone being shattered as Dunaway went down with a roar of pain and surprise.

The INS agents looked at each other, realizing what had just transpired, and the ramifications that would soon follow. It would be messy. Cops shooting cops always was. They started to move toward the two men now stretched out on the sidewalk, becoming aware of porch lights coming on in the neighborhood; doors were opened and quickly slammed shut.

They had advanced only a couple of steps when two streaks of light flashed by them. Feeling a brush against their elbows, they then felt nothing at all, as they crumpled to the bare ground face first, their weapons clasped firmly in their hands.

Mickey and Jan had joined Heidi and Blanche at the front door, observing the mayhem in front of the house. They watched as Rhett and Jed left the two agents, moving toward the sidewalk to check on the men lying there. Rhett leaned to touch the chief, who suddenly fell silent. The siblings still had no clue about the identity of the wounded men, and would not have the chance to find out until later. Blanche beckoned them back inside and closed the door, as she and Heidi stood facing them.

"Mickey, you and Janice must go now. It is important that no one becomes aware you were here to witness this," Blanche said, leaning in for emphasis.

Confused, and more than a little frightened, Mickey responded, "But why? We didn't do anything."

"Of course you didn't. But people, including the authorities, will want to know why you were here, and what have you seen. That will lead to questions about our identity, and what we have shared with you." Blanche looked toward the street. "I can hear the sirens of the police vehicles now. You don't have long before they arrive. You and Janice must hurry."

Before Mickey or Janice could say anything, Rhett and Jed entered, closing the door behind them.

"Blanche is correct, Mickey," Rhett said. "They will be here within minutes. Neither they, nor the two agents, should have any knowledge of your presence. If they do, you will be interrogated until you will have no choice but to expose our existence, and what you know about us."

"But what about the men out there?" Janice pleaded, pointing. "Won't they know?"

"Both the two agents and the man who was the shooter will be fine. He is the man who confronted you on our property not long ago ago. His wounds are not fatal. Unfortunately, the same cannot be said for the other man. I'm sorry, but he is lost. The other three will be unconscious for ten minutes, so you must go."

"But," said Mickey, desperation showing, "you'll still be here. They'll interrogate you."

Heidi stepped forward. "No, Mickey. We are leaving. We have no choice."

"Leaving? Will you be coming back?" A sudden jolt to his heart gave him the answer before Heidi had a chance to respond.

"No, Mickey. It will never be safe for us here now. I told you before we may have to re-locate at some point."

"But . . ." He stopped, unable to speak without choking up. It was suddenly clear to him why the equipment had been left in the transport.

They knew this might happen. It was also becoming clear that it was his battle with Chief Dunaway, who now lay wounded on the sidewalk, that was causing them to leave. It was his fault they were going. He felt a kind of stomach-twisting sickness and shame he had never felt.

It was left to Janice to speak for him, to ask the things he couldn't. "Will we ever see you again? Any of you?"

"That will likely be impossible, Janice. As with most things we have shared with you, it is for your benefit as well as ours," Blanche said.

"Will we be able to know where you go?" The girl was close to tears now herself.

"We also do not think that will be likely. It will be necessary to guard our new location." Blanche was holding the communication device Janice had been making calls from earlier. She extracted a small square from somewhere within the device, closed it again, then handed the device to Janice. "You may keep this. I have de-activated it, so it cannot be used for communication, but perhaps someday, you may want to share this with Earth's scientists, as proof of our time here. They will be able to examine it and identify it for what it is. But please, wait for some time before showing it to anyone."

The teary-eyed girl took it, clutching it to her chest. "We will, and thanks."

"It has been our pleasure, Janice and Mickey. You are both good humans. If more on your planet were as you are, our secrecy may not have been necessary. But for now, you must go. You may want to exit through the rear door to avoid being seen. Hurry now."

As the siblings started to leave, Mickey was overcome with the finality of the situation, his emotions bubbling over. He reached for Heidi, his friend and classmate, hugging her as gently as possible. Feeling the faint tingle of the electro-magnetic field in her alien body, he felt as if he could squeeze right through her human form, right down to her natural state. He quickly let go, afraid he might harm her.

"Goodbye, Heidi. I'll never forget you," he sobbed.

"Goodbye to you too, Mickey." Heidi's words were the same tone, devoid of feeling, as they had been the first time she had spoken to him weeks earlier.

With that the siblings made their way to the rear door, doing their best to find their way while dealing with blurry eyes streaked with tears.

They circled through the back yard, then stayed behind the next two houses, knowing the occupants were probably toward the front of the houses, checking out the action. When they were clear, they ran hurriedly across Louisiana Street and looked for cover.

"Look," Janice whispered, pointing to a bed of thick bushes. "We can watch from here." They crouched down, just as two police cars turned the corner at Sabine, followed by an ems ambulance. The scene in the front yard of 710 Louisiana Street would soon be chaotic. No one knew exactly what had transpired. There were two men on the sidewalk, and two others rolling in the yard. The INS agents were slowly coming to life; they were shaking their heads, trying to get their bearings. Four officers climbed from the Cypress Springs Police Department patrol cars, heading first for the sidewalk.

"Oh crap!" one yelled. "It's the Chief!"

"Who are those guys?" another asked, pointing to the men climbing to their feet. Officer Brad Sweat was in the group, recognizing the two agents. “Crowley! What the hell are you two doing here?”

Crowley ignored the question. Followed closely by Medlin, he walked on rubbery legs in the direction of Dunaway, who slowly appeared to be regaining consciousness. “What the hell were you thinkin’! Have you gone loco, Dunaway?” Crowley bellowed,

One of the other officers started to draw his weapon, approaching the agent.

“Who the hell are you?! Can’t you see he’s wounded? Who are these idiots, Sweat?”

But before he could answer, another officer, along with an ems technician, were checking out the other man on the concrete. “This one’s gone. No pulse,” the officer announced. “Hey, I think I recognize the guy. Pretty sure it’s Jim Robinson. Did some work at our building last year.”

Agent Medlin was the only law enforcement person who seemed to be keeping his wits about him. “Let’s get the Chief to a hospital. Get him taken care of. We can finish our pissing contest later.” He held out his badge for the locals to see.

Amidst the yelling and name calling, a smaller form in comparison to the adults gathered around was approaching. He was stopped by an officer when he drew too close.

“Sorry, son. You can’t be here.”

The boy pointed. “I think that’s my dad.”

The officer followed the gaze. “Well, I’m sorry, son. What’s your name?”

“Butch Robinson.”

The patrolman took the boy by the shoulder, guiding him away from the scene. "Well, son, I'm afraid he didn't make it. Is your dad's name Jim Robinson?"

"Yes," Butch answered, showing no emotion. But there were signs of fresh tears streaking down his cheeks. "He hurt my mom pretty bad. I think she needs help."

Everyone on the site heard the boy, all wondering in unison how the night could get any stranger. The officer grabbed one of the emt's and together they followed the boy to his house.

Able to hear enough of the conversations in front of the house, Mickey now sunk lower behind the shrubbery. A new layer of guilt was being absorbed. Butch's father dead, and his mother probably in bad shape. All due to his wandering eyes and the desire to one-up his friends. Jan would disagree with that, he was sure, but he knew the truth. He put his head in his hands.

Janice interrupted his angst, pulling him up by his shoulder, whispering, "Look, Mick!"

He reluctantly followed her gaze, in time to see four spinning swirls of light exiting the house from the back door. The neon green forms circled the roof of the house once, and then vanished through the trees, invisible to the group of men in the front yard, too busy trying to figure out their crime scene to notice.

Mickey, now inconsolable, watched helplessly as they left his blurred field of vision. Real tears were streaming now; he had to bury his face on Jan's shoulder to keep from being heard. She sobbed along with him. The best friends he had ever had, and would probably ever hope to have, gone.

They had come into his life in an amazing vehicle, dragging fire behind them as they found his part of Texas; now they were leaving, with nothing remaining to prove to anyone they had ever existed on Earth, other than Mickey and Janice Foreman, two school kids of Cypress Springs, Texas.

EPILOGUE

February, 1992

Over the span of years, Mickey discovered he been more right than wrong in regard to future friendships. There were some who he had grown close to, mostly on the female side, and they to him, or so he wanted to think. But when it came to unconditional acceptance, without fear of judgments being applied, and very little being asked in return, nothing had come anywhere near those few short weeks. What the residents of 710 Louisiana Street had shown him and given him in a space of less than two months was a remarkable part of the wonder he had experienced. His sister had voiced the same to him on many occasions. Their only price had been to keep what they had learned to themselves. But that had proven to be a steep price; a price they continued to pay.

It had all come to a crashing, devastating, sudden end that violent night, an end that had changed the lives of several people, including his own. Following the death of his father, Butch and his mother had moved away from Cypress Springs as soon as Jane Robinson had recovered from the damage her husband had done to her. The ribs healed themselves within weeks, but the crushed eye socket took much longer. She had a sister in Henderson, Texas, and knowing she couldn't stand the shame in her small town of her affair with Winston Dunaway, she packed Butch up and went to live with her.

Chief Dunaway fared worse, at least from his ego-driven viewpoint. While he was recovering from his gunshot wounds, the ram-rod straight mayor terminated his job. It wasn't so much the fact he had killed an unarmed citizen, a charge that was ultimately reduced to manslaughter, but when the reason for the confrontation came to light, his career was doomed. His insistence that the former residents of 710 Louisiana Street were somehow illegal aliens, who possessed strange powers, made no difference, since it had no bearing on his crime. Not that it mattered; no one was listening to him any longer. The INS agents, who had been placed on paid administrative leave for discharging their weapons, refused to back up his story, choosing never to answer questions about that night. The case that brought them to Cypress Springs, which wasn't an official case at all, was permanently closed. And, of course, the mysterious disappearance of the "Jones" family didn't help the disgraced chief's cause either, since there was no one left to take statements from. Mickey and Janice's involvement in the night's events remained known only to them. With incredible speed, even for the amazing extraterrestrials that they were, they had reduced the monitors in the rear room, plus any other evidence of their time in Cypress Springs to ruble, leaving smatterings of hard to identify pieces of metal.

Mickey's life for an extended period of time had been in a deep hole following that night. He grew so despondent that his confused mother had considered engaging heavy counseling for her son. It was Janice who had urged her not to, finally managing to convince Jessica Foreman that her brother's depressive state was due mostly to his continued self-blame for causing the tragedy involving Butch's family. The siblings had explained to her soon after the shootings the truth of Mickey's knowledge of the affair, and his blabbing it to his friends. Another reason for his moods, according to Jan, was that he and Kristi had broken off whatever they had (which was true). Jessica accepted those reasons, but none-the-less kept a close watch on her son. Of course, the real push into his personal deep well was the sudden departure of Heidi and her alien family, which was the one factor neither he nor Jan would ever mention. If they had, Jessica would have thought both of her kids needed help after telling such a ridiculous, fantastical story. Much to their relief, she never dwelled too deeply into the Jones' sudden departure.

And they hadn't told anyone, even after all the passage of time—fourteen years in all. It was such an unbelievable secret that they amazed even themselves by keeping it.

Those were his final trips for now into his memory banks as he noticed the gate opening and the flight he had been waiting for emptying out. As he had done an hour earlier, she had let most passengers deplane before making her appearance. After all, they were family, and like-brother-like sister.

He embraced Janice with a bear-like hug; it had been too long. They pecked each other's cheek, then stood back to examine each other.

"I'd forgotten how tall you were now," she said, having to gaze upward.

It was true. He had shot up six inches over his senior year in Cypress Springs High and his freshman year at North Texas State, putting him at six-three. Ironically, when his growth spurt had ended, he no longer wanted the western boots he had looked forward to. That footwear wasn't cool in college, not even in Texas—except perhaps at Texas A & M. Jan had leveled off at an attractive five foot-nine, which looked appealing with her still-lithe body. Her hair was now just like her mother's had been in younger days. She looked great, Mickey thought.

"After years of being a runt, I think I deserved a little boost. Does help some in my busy, busy social life."

"I'll bet. How many starlets have you bedded lately?" She gave a friendly punch to his belly.

"None that I'm at liberty to talk about," he answered with a laugh.

God, how he enjoyed these infrequent meetings with her. It was one more result of those incredible weeks back in northeast Texas, including the way it had ended. It had drawn them closer. Before, they had only been brother and sister, taking things a day at a time with their overwhelmed mother in rural Texas. Since that time, they had shared the bonds of secrecy and experiences they were pretty sure no siblings ever had. The distance between their everyday lives hadn't affected those bonds.

"You figured out where the baggage claim area is?" she asked.

"Right. Let's head there, then we'll pick up our rental." Clasping his brief case under one arm, he offered to carry her overnight bag, but she waved him off.

Stepping off the escalator on the level below, where baggage claim was located, they found the carousels for their respective flights adjacent to each other, but there would be a wait involved before Janice's luggage would roll out.

As they stood and chatted, Janice addressed the main issue of the reunion. "Well, are you ready for this?"

He shook his head slowly. "I'm really not sure. I've gone back and forth. But I'm going to find out."

"*We're* going to find out," she corrected her brother. "Wouldn't've come if I didn't think we'd follow through." She people-gazed for a second, then asked, "How much time are you going to allow for this. I mean, when are you going to do what you were actually sent to do?"

"Good question. I thought we might talk about it over dinner tonight. It's on me, by the way."

"Great. Where we going?"

"Cheapest Tex-Mex place I can find. Phoenix has a ton of 'em."

Her question had indeed been a good one. The project he was being paid to perform in Arizona was finding filming locations in the Arizona desert for a movie he was involved in. Although they had ebbed and flowed over the years, his dreams of having some role in the glamour of filmdom had come to a degree of fruition. Through a series of fortunate moves and connections made, first at North Texas State, then his final two years at U. T, following his then-girlfriend to enroll in the Austin university's new Film and Music curriculum, he had made his way to Hollywood, even though his residence was currently in a dingy apartment in Inglewood. He shared the unit with a guy who had been a year ahead of him at U. T., and who had hooked Mickey up with a medium sized production company. His title on the current project, which was on paper only, was Assistant to the Director, but was actually a flunky position. He did whatever he was asked to do to enable him to say he "worked in the film industry".

As with most things in life, showbiz wasn't all it appeared to be from the outside. Being in and around studios was exciting, as well as catching glimpses of stars he had admired for years—actually meeting a few. But he was sure the paranoia, back-biting and narcissism would eventually drive him away from it. Either that, or become one of them, which wasn't a palatable choice.

When he became aware of the need for a set location near Phoenix, he had camped outside the director's office until he was given the assignment. But now that he was here, it would have to wait a while until other matters were addressed.

Later that night, the cantina selected, proving to be a popular local spot, was named The Barrio, and to Mickey's surprise, wasn't as inexpensive as he had thought. But he had committed himself, and with an expense account to back him up, he sucked it up, happy to treat his loyal sister. They both remarked on the memories of their nights with their mom in Senor Pepe's, which was now being managed by Roby Contreras.

Nursing frozen margaritas, he laid out his proposed schedule. "I thought we might head north around nine in the morning. I see our drive at around two-and-a-half hours. I'm pretty sure whether we're successful or not, one way or the other we'll be back in Phoenix sometime after dark." He took a sip, pinching the bridge of his nose to ward off brain freeze. "I'll start seriously looking for sites the next morning."

Janice stared across the heavy, thick wooden table for a second. "How about if I ride along with you that day. Might be fun helping you find a spot I might see on the big screen some day."

"I don't mind, but it'll probably be pretty boring, ridin' through desert after desert. I'm not even sure how to go about it. This is the first time I've done this, remember."

"Well," she pointed out, "I've gotta do something that day. It'll be Saturday, and my flight back to DFW doesn't leave until Sunday afternoon." She paused and gave that evil, teasing smile he had seen a thousand times. "Wouldn't it be funny if I actually picked out the right location?"

Mickey scratched the four-day old stubble he now maintained on his face, which was the style among his peers in L. A. After giving her request some thought, he said, "You've likely got a better chance than I do. So, sure, why not?"

With plans for the next two days settled, they relaxed, stuffing themselves with chips, salsa, and a huge order of chicken fajitas they shared. It was a night that only two people who cared deeply for each other, and had shared so much, and hoped to share more the following day, could fully appreciate.

The first letter had arrived roughly nine months after they had disappeared into that cold, clear, deadly November night. Except, it really wasn't a letter at all; it was simply an envelope, with nothing inside. It was addressed to Mickey Foreman, at Jessica Foreman's house. The only clue was the return address, which consisted of only:

JONES

PALATKA, FLORIDA

No street address or any other markings other than a post mark indicating the date and post office from which it was sent. Fortunately, Jan was home, enjoying her last week of summer before school started back up, and was able to collect the letter before Jessica arrived, which had saved an awkward explanation about what it meant. She and her brother met in her room as they had before, wondering what the sudden communication meant. More elated than surprised, Mickey had no idea. It was contradictory to the alien family's lack of anything related to sentimental feelings of the past, plus they had indicated any notification of their new location was unlikely.

Other than quickly locating the town on a Florida map, they did their best to put it out of their thoughts, hoping for more envelopes yet to come. Mickey copied the name of the town on a piece of paper, hid it deep in his backpack, then tore the envelope into shreds and flushed them.

The next came two years later, addressed similarly, only this time from a town by the name of Borrego Springs, California. So much time had elapsed, the euphoria of the siblings wasn't as great this time, but still encouraging enough to excite them. Jan was convinced they would continue to stay in touch, and might eventually give more clues as to their exact addresses. She would prove to be correct.

With the next envelope, which came just over two years after the second, there was a change. This time, the return address was a town in southeastern Utah. Mickey was ready to tear the envelope apart before shredding it, when a small piece of paper fell out. On it was an actual street address in the town of Moab, Utah. After the excitement the siblings shared had once more worn off, Jan pointed out that since Mickey was now fifteen, going on sixteen, and she was nearing eighteen, they needed to get a message to Heidi and her group. Her grades were hovering toward perfection, and she was sure to receive several scholarships. College was definitely in her future, and Mick's, hopefully, a few years later. There would come a time when a letter such as this arriving at their home would be confusing to their mom; she would ask questions.

"Mick," Jan had said, "you need to write a letter to Heidi, explaining that we both will more than likely have new mailing addresses in the next few years. Tell her we'll send my college mailing address when I get one, and for them to let us know if they move to a new community before they receive my address. That way, hopefully Mom won't ever open one of their envelopes, which we definitely don't want to have to explain."

"Good idea," Mickey had replied. "I'll do it first chance."

Jan had thought about it for a second, then added, "I'll bet they had already figured out we'd have to do that, knowing them, and how they think ahead."

That plan worked well, and as soon as Jan settled in to the Southern Methodist University campus in north Dallas, she sent her address to the alien family. More updates in both directions were required over the next decade, until an address in Sedona, Arizona, arrived in Jan's mail box in Plano, Texas, in late 1991. By that time, with an MBA in hand from SMU, she was on a relatively fast-track as an HR executive with a real estate development company in the booming metroplex.

The result of that sporadic fourteen years of correspondences found Mickey and Janice heading north from Phoenix on I-17 on Friday morning. They got a late start, deciding to let the morning rush abate, putting them on the road shortly before ten a. m. Exactly why they placed such importance on their mission was hard for either of them to put in words. It had something to do with wanting to seek verification for what had seemed so unworldly during the fall of 1978. They also had agreed, even while planning their reunion, that this was likely the only chance they may ever have to accomplish whatever the outcome would be. Then there was also the underlying reason—they really wanted to see the friendly, yet stoic, brilliant group. Another agreement shared was that the letters sent by the group from another galaxy was their way of inviting the two Earthlings to find them—for whatever reason, they could only speculate.

Arriving in Sedona shortly before one p. m., they were immediately struck by the startling beauty of the small town. Leaving the interstate, they had traveled for some time on twisting, nondescript county highways, until rounding a final curve, and the road emptied into a scene of striking red cliffs of varying shapes and sizes. The setting of the jewel of a town was so breathtaking, they nearly lost track for a few minutes of why they were there.

"This is unbelievable," said Jan, awestruck by the natural scenery. "I can't believe I haven't heard about Sedona before."

"Right. I can see why they chose this place." Mickey had to concentrate his attention on the wheel, while looking in all directions. "I could definitely live here."

"Maybe a choice for your movie scenes?" Jan looked at him hopefully.

"I would say so, except my boss wants something closer to Phoenix. And a lot rougher than this. This is way too scenic."

A stop at the town welcome center was needed to locate the street, which turned out to be a road, a road from which the latest "Jones" letter had originated. It then took most of an hour to find the address, even after finding the correct road. It exited the town on the opposite side they had entered on, twisting through red-splashed canyons and shallower barrancas, a trip that brought back distant, but clear memories of those blind-folded rides out to the transport hidden in the woods outside Cypress Springs. Most houses on the narrow road sat back at least twenty yards or so, hidden behind cacti and other vegetation, with numbers on mailboxes roadside nearly obliterated. Finally, after missing it the first time by, they found the correct house number.

Pulling into the driveway, Mickey rounded a curve about twenty feet in, then came to an abrupt stop when he saw the house only yards away. Pulling into the short driveway, the siblings looked at each other, taking deep breaths. It took a few seconds to build up enough momentum to step from the rental sedan; they finally climbed out tentatively. It was a small, stucco-walled structure, appearing to be roughly the same size as the house back in Texas. Large wood beams framed the corners, as well as the front doorway. They approached slowly.

It should have been no shock to either of them when the door swung open before they reached it, causing them to halt and take a step back.

"Mickey, Janice. How nice to see you." There was actually a hint of a smile showing on the greeter's face, actually more recognition than anything.

The fact the "being" they had known as Heidi Jones had anticipated their arrival also shouldn't have been a surprise; they recalled that ability, but there was more to it than that. Standing before them was definitely the same "girl" they had known, but she had changed. A few inches taller, her pale skin from before was now slightly bronzed, similar to a human who had stayed out in the sun a little too long. The fine fake hair was bleached to an even whiter shade than before. The Arizona sun had taken a toll. The one thing that hadn't changed was the voice; still firm but somehow soft, calm in tone.

Mickey and Janice leaned into each other for a second, the emotion of the moment finally being realized was taking hold. It was Janice who managed to say, "It's great seeing you again too, Heidi."

They gazed around the interior of the Spanish-styled living room, which was smaller than the living room/dining room combo from earlier. Sunlight filtered in through slatted windows, displaying Spartan furnishings next to beige walls, with potted plants spread around. The effect was definitely in line with what they had seen so far in Sedona; bright, and decidedly southwest casual.

Mickey peered behind Heidi; they were all there: Blanche, Rhett, and Jed, standing and staring at them, also with odd grins displayed. Jed's floppy hat had been replaced by a straw hat, similar to the type he had seen some golfers wear to ward off the rays of the sun. The hair of the other two, still wig material of some variety, was bleached to a shade similar to Heidi's. Their skin layers were also tanned.

"It's great seeing all of you again," he said. "It's been a long, long time."

"Yes," replied Rhett. "it has. However, as we discussed during that time, your measure of the passing of time is different than ours. Plus, we have been kept busy moving several times."

"So we've noticed. But this place seems special. Sedona is amazing."

"Yes, it is," Blanche agreed. "A previous location we occupied for a while in the state of Utah was similarly nice, but not as dramatic as this. Another advantage to this area is that the residents are somewhat offbeat, being an art-conscience community. That enables us to fit in rather well."

"By the way," Heidi interjected, "we are now using new names. However, it will be fine with us if you wish to address us as you knew us, since our Earth names are not important to us."

Janice couldn't resist; she had to know. "Okay, but do you mind telling us what you are calling yourselves here?"

"Certainly. Similar to our time in your community, as your wise mother uncovered, we are again using designations from one of your former popular television shows. I am Gloria, Blanche is now Edith, Rhett is Archie, and Jed is George. Our last names are now Smith."

"Yes!" Janice said. "All in The Family" characters." She paused, her brow furrowed. "But who is George?"

"I know," said Mickey. "George Jefferson. Their neighbor. He had his own show later."

"In the flesh . . .well, in the skin," Jed said, spreading his arms.

That comment was so out of character compared to fourteen years earlier, it gave the siblings pause. But they couldn't resist a laugh.

Mickey was busy doing another mental comparison; in one way things were just as they had been back in Cypress Springs, but from another perspective, things were different—more informal. Had they changed? adjusted to a more human way of looking at things?

"How is your mother, Mickey?" Blanche interrupted his thoughts.

Mickey and Janice brought them up-to-date on Jessica, explaining that she had recently married her long-time boss. John Montgomery had suffered a mild stroke three years earlier, caused by the stress of his work, according to his doctor, and seconded by Jessica. He had mellowed during recovery, softening toward the fact Jessica Foreman was the mother of two kids who meant the world to her. Before the knot was tied, there had been a group meeting involving Mickey and Janice during trips home, and the engaged pair, with assurances given that the happiness of their mother was the most important part of the merger, and that her offspring would always be welcome. The couple was now residing in Montgomery's comfortable Mt. Pleasant home. Now in his late fifties, he had slacked off, delegating responsibilities to younger employees who he had plans to sell his company to eventually.

"Her life is much easier now," Janice concluded.

"That sounds good. We are happy to hear that. We all regret we did not have a chance to meet her."

The alien group, now referring to themselves as Smiths, then spent time describing their journeys over the past fourteen years, and the places they had settled. When that history had been completed, the siblings took quick glances at each other; Janice nodded her okay.

"We were surprised when you started sending us your locations," Mickey said. "But glad, of course. When you left that night, you didn't give us much hope."

Rhett, aka Archie, leaned in from the chair he was seated in. "Yes. We must explain that to you. Do you two have time to listen?"

"Of course. All the time that you need." The siblings leaned forward on the rattan sofa they were seated on. They had come this far, and were anxious to learn more from the alien friends they never dreamed they would see again. During the drive to Sedona, they had agreed it wasn't at all critical to arrive back in Phoenix that night, especially if something better cropped up; something such as this.

"It was actually a test," Rhett began. "We were hoping to discover if you had the desire to remain in contact with us."

"Wow," Janice said. "There was no question about that. But what do you mean? Why a test?"

The alien family shared their own glances, and with nods all around, Rhett continued.

"When we left Cypress Springs, it was necessary for us to again place our transport in orbit. While there, we communicated with others from our world. We began to have discussions as to our future plans while we were on your planet. Opinions were offered that we had to make a choice. Either return to our world and end the experiment, or begin the process of assimilating ourselves into yours. Our preference, which was thankfully accepted, was the latter. We simply had too much time invested to put it all behind us."

Mickey shared a brief look of excitement with his sister. This was sounding promising.

"And if we were going to adjust to life on Earth, we believe that eventually we must identify who we are to Earth's inhabitants, in hopes we can share our science and technology. We believe we have a lot to offer."

"There's no question about that," Mickey said.

"However, if we are to do that, which may take some time to

accomplish correctly, we will require verification by someone on your planct as to who we are. In other words, someone to introduce us when the time comes, and to vouch for the fact we mean no harm and are not threats to your way of life. We are convinced you two can serve that role, but only if you choose to. Your searching us out has given us evidence that you at least do want to remain in contact with us."

Mickey felt as if he might explode with excitement from the words spoken. He gauged Jan's expression, which showed she shared his feelings, but to a lesser degree.

"Before you make your decision," Rhett continued, "you must be aware of two things. It will take some time for us to make a final decision as to the proper time for us to begin that process. We have begun attempting to build some of the feelings and what you refer to as emotions into our daily life, which you may have noticed."

"Yes," Mickey said. "I thought you were all a little . . . well, different." Laid back, was the best definition. The fact these beings could pull that adjustment off didn't surprise him for a second.

Janice nodded agreement, smiling broadly.

"It is simply observing the reactions of humans to various situations and comments, and matching ours to theirs. However, we are certain there is more for us to learn.

"The second point to keep in mind before you commit, and this is much more important, as far as you are concerned, once our full assimilation is put into motion, and you verify our identity and what we have shared with you, other humans may be suspect of not only us, but you as well. There will be countless questions for you to answer about us, and why you have chosen not to mention our presence in your world before. Our motives for being on your planet will be questioned, perhaps sometimes even violently. For these reasons, we do not expect your answer at this time. Think it over for as long as you need."

The weight of what Rhett had laid out seemed to strike Janice more clearly than her younger brother. She continued to be the more introspective of the two, wanting to reel in Mickey's excitement.

"Rhett's right, Mick. We need to decide if we want to be bombarded with those doubts, as exciting as our role would be."

Blanche leaned in. "Have you kept the communication device we gave you before departing your community?"

"Yes. I have it in a safe deposit box at my bank. Didn't want it lying around my place for others to see and for me to have to explain." Janice's face brightened. "That would be solid proof, wouldn't it, for both you, and Mick and I."

"Yes indeed," Rhett said, "but even with that you must be certain you want to be . . . shall we say . . . our emissaries when the time comes. And you will have plenty of time, perhaps even years, to prepare."

"Wait," Mickey interjected. "I have a bag phone my boss told me to rent; it's in our car right now. It's like a mobile phone, or something. Is that sorta like the device Jan has now; the one you gave her when you left?"

"Yes," Rhett answered. "I believe that is a prototype of sorts for what you will see in Earth's future. We predict that within ten years or so, in Earth years, the technology will produce devices approaching what we left with you. However, the one we gave to you will still have more advanced features and materials to distinguish it from the ones humans can devise."

"The fact wc have time to consider this does sound better," Janice said, getting back to the proposal made. "We'll discuss it every time we talk from now on." She paused, then looked at the alien group, one by one. "For my part, the confidence you are placing in Mick and me is an enormous honor, to say the least. When the time comes, if we agree to perform that role, I hope we can live up to the faith you are entrusting us with." She appeared to her brother to be putting her HR experience into action, using carefully chosen words.

"You have our full confidence," Jed said. The others nodded as they smiled, evidently testing out the feeling of happiness still under development.

The six of them then talked of many other things for a while, until Mickey decided to change the subject by asking a question, somewhat tentatively. "Is your pod, rather, transport, hidden in the desert somewhere nearby?"

Jed, aka Archie, practiced smiling again. "Of course. I thought you would never ask. Only farther out, compared to Cypress Springs, away from town."

Mickey paused for a moment, then asked, "Well then, is there any chance . . .?"

"Are you suggesting another excursion?" Jed asked, his grin broadening.

"Well, yes. If you're willing. We have a car, so may as well put it to use."

"Certainly. You have seen the Grand Canyon, and Las Vegas has grown, too much in our opinion, and more difficult to hover above with all the new construction." The alien in the straw hat, now calling himself, George, looked at Heidi. "However, we have a new destination in mind."

She smiled back.

An hour later, as dusk began to show promise in the light desert skies, and the red rocks took on darker shades, the same four travelers from years ago wedded the top to the bottom of the new, improved, more spectacular version of the transport craft, and lifted off. The two humans were informed the speed of the pod had also been drastically enhanced, so, "Buckle up".

As before, meaning fourteen years earlier, Blanche and Rhett stayed home, monitoring the action, anxious to see the reactions of the human brother and sister who had accepted them into their lives years earlier.

"Next stop," Jed announced, "arriving in fifteen minutes, San Francisco Bay."

And they were off.

Author's Notes

As the cover page indicates, this is a novel; a work of fiction, at least, as far as I am aware. The debate over whether life exists beyond the Milky Way, in Adromeda, the nearest galaxy outside our own, or in any of the trillion or so galaxies scientists estimate are out there, will continue until proof is provided on either side. However, that proof will almost certainly have to come from the positive side, since, similar to the phrase, *you can't prove a negative*, there will unlikely ever be any evidence put forward that beings *cannot* exist anywhere. I would label myself a skeptic at this point, but I can dream, and have done so through the words in my story. And, as long as there are adventurers like Elon Musk attempting to find proof, and astronomers like Neil deGrasse Tyson to chronicle it, I will never say never.

I would like to give thanks to Al Esper, of Al Esper Graphic Designs, for his expert cover design, Maddie Lamb, who initially suggested the theme for Al's final design, and Mike Hartley, who saved me from abandoning this and past projects with his technical skills. I would also like to thank Isabel Navarette, Maddison Machart, and Dawn Marinich, along with the residents of Morningside of Belmont, who aided and abetted me in my writing efforts. I love them all.